# WHY WE
# MAKE
# MISTAKES

. . . . . . . .

# WHY WE MAKE MISTAKES

· · · · · · · ·

*How We Look Without Seeing,*

*Forget Things in Seconds,*

*and Are All Pretty Sure*

*We Are Way Above Average*

· · · · · · · ·

## Joseph T. Hallinan

Broadway Books
New York

Published in the United States by Broadway Books, an imprint of
the Crown Publishing Group, a division of Random House, Inc.,
New York.
www.crownpublishing.com

BROADWAY BOOKS and the Broadway Books colophon are trademarks
of Random House, Inc.

Originally published in hardcover in slightly different form in the United
States by Broadway Books, New York, in 2009.

The author and publisher gratefully acknowledge and credit the following for
the right to reprint material in this book:

*Turning the Tables* illustration: *Turning the Tables* by Roger Shepard from the
book *Mind Sights* by Roger Shepard. Copyright © 1990 by Roger N. Shepard.
Reprinted by permission of Henry Holt and Company, LLC.

*Pennies* illustration based on: *Cognitive Psychology* 11, Raymond S. Nickerson
and Marilyn Jager Adams, "Long-Term Memory for a Common Object,"
pp. 287–307, © 1979, with permission from Elsevier and Raymond S. Nickerson.

Library of Congress Cataloging-in-Publication Data
Hallinan, Joseph T.
    Why we make mistakes / Joseph T. Hallinan.
        p.    cm.
    Includes bibliographical references and index. 1. Failure (Psychology)
2. Errors.  I. Title.
BF575.F14H35 2009
153—dc22            2008030818

ISBN 978-0-7679-2806-9

Printed in the United States of America

Book design by Donna Sinisgalli
Illustrations by Jackie Aher

10

*For*
Jack

*for*
Kate

*for*
Anne

but most of all
*for*
Pam

To Howard K. Hess,
*the best friend a man could have*

Mike's belief, and I subscribe to it myself, is that at the exact moment any decision *seems* to be being made, it's usually long after the real decision was actually made—like light we see emitted from stars. Which means we usually make up our minds about important things far too soon and usually with poor information. But we then convince ourselves we *haven't* done that because (a) we know it's boneheaded, and no one wants to be accused of boneheaded-ness; (b) we've ignored our vital needs and don't like to think about them; (c) deciding but believing we haven't decided gives us a secret from ourselves that's too delicious not to keep. In other words, it makes us happy to bullshit ourselves.

—Richard Ford, *The Lay of the Land*

# Contents

# WHY WE
# MAKE
# MISTAKES

## Why *Do* We Make Mistakes?
## Because . . .

There are *all kinds* of mistakes. There's real estate you should have bought and people you shouldn't have married. There's the stock that tanked, and the job that didn't work out, and that misguided attempt to save a few bucks by giving yourself a haircut.

And then there are the errors of other people. As a newspaper reporter for more than two decades, I have made a small (and arguably perverse) hobby of collecting stories about these, tearing out the more memorable ones and tucking them into a manila folder I had labeled "Mistakes."

My favorite was torn from page 34 of my hometown paper, the *Chicago Sun-Times.* The story involved an incident that occurred a few years ago in the village of St. Brides, in South Wales. According to the Associated Press, a mob of vigilantes attacked and vandalized the office of a prominent children's doctor there.

Why attack the office of a prominent children's doctor?

Because, according to police, the vigilantes had confused the word "pediatrician" with the word "pedophile."

The doctor involved, Yvette Cloete, was forced to flee her home, which had been spray painted with the word "paedo"—an abbreviation for the British spelling of "paedophile." Afterward, she gave an interview to the local paper.

"I suppose," said Dr. Cloete, "I'm really a victim of ignorance."

## To Err Is 90 Percent Human

She was, of course. And so are we. We all know the cliché "To err is human." And this is true enough. When something goes wrong, the cause is overwhelmingly attributed to human error: airplane crashes (70 percent), car wrecks (90 percent), workplace accidents (also 90 percent). You name it, and humans are usually to blame. And once a human is blamed, the inquiry usually stops there. But it shouldn't—at least not if we want to eliminate the error.

In many cases, our mistakes are not our fault, at least not entirely. For we are all afflicted with certain systemic biases in the way we see, remember, and perceive the world around us, and these biases make us prone to commit certain kinds of errors. Right-handed people, for instance, tend to turn right when entering a building, even though that may not afford the best route to take. And most of us, whether left- or right-handed, show an inordinate preference for the number 7 and the color blue. We are also so swayed by our initial impressions of things that we are reluctant to change our first answer on a test; yet many studies have shown we would be better off if we did exactly this.

> Most of us show an inordinate preference for the number 7 and the color blue.

Our expectations can shape the way we see the world and, often, the way we act in it as well. In one case, people encountered an unknown man and were later told his occupation. When they were told that the man was a truck driver, they said he weighed more; when they were told he was a dancer, they said he weighed less. In another case, half the people in a restaurant were told their complimentary glass of cabernet sauvignon that night came from California; the other half were told their wine came from North

Dakota. Not only did the North Dakota group eat less of their meals, but they headed for the doors more quickly. Even presumably stolid people, like farmers, show the same propensity. Farmers who believe in global warming, for instance, have been shown to remember temperatures as being warmer than those recorded in statistical tables. And what about farmers who do *not* believe in global warming? They remembered temperatures that were colder than those in the record books.

What's important about these examples is not that we think a trucker is fatter than a dancer or that temperatures are warmer than they used to be (unless, of course, you like to bet on these kinds of things). What's important is that these effects occur largely outside of our consciousness; we're biased—we just don't know we're biased. Some of these tendencies are so strong that even when we do know about them, we find it hard to correct for them. A practical example involves the power of first impressions. Nearly eighty years of research on answer changing shows that most answer changes are from wrong to right, and that most people who change their answers usually improve their test scores. One comprehensive review examined thirty-three studies of answer changing; in not one were test takers hurt, on average, by changing their answers. And yet, even after students are told of these results, they *still* tend to stick with their first answers. Investors, by the way, show the same tendencies when it comes to stocks. Even after learning that their reason for picking a stock might be wrong, they still tended to stick with their initial choice 70 percent of the time.

## We Take the Good with the Bad

Biases like these seem to be deeply ingrained in us. Many of the qualities that allow us to do so many things so well contain flip sides that predispose us to error. We happen to be very good, for instance, at quickly sizing up a situation. Within a tenth of a second or so after looking at a scene, we are usually able to extract its meaning, or gist. The price we pay for this rapid-fire analysis is that we miss a lot of details. Where the problem comes in is that we don't *think* we've missed anything: we think we've seen it all. But we haven't. An obvious example comes from Hollywood. Movies are typically made on film composed of individual frames that are exposed at a rate of twenty-four frames per second. But when the film is projected onto a screen, we don't see still pictures; we see *motion* pictures. This is a "good" error, of course, and we don't mind making it; in fact, we usually enjoy it. But a similar visual error, committed by the doctor who looks at our X-rays for signs of cancer or the airport security agent who looks for bombs in our luggage, can have deadly consequences. And, as we'll see, they miss quite a lot of what they're looking for.

## The World Around Us Isn't Helping

Simply put, most of us aren't wired the way we think we're wired. But much of the world around us is designed as if we were. We are asked, for instance, to memorize countless passwords, PINs, and user names. Yet our memory for this type of information is lousy. In one test, 30 percent of people had forgotten their passwords after just one week. In another test, after three months, at least 65 percent of passwords were forgotten.

> In one test, 30 percent of people forgot their passwords after one week; after three months, at least 65 percent were forgotten.

Or we are pushed to our limits by lives that demand multitasking, even though the number of things we can do at once is very limited. Exactly how many things depends in part on the type of thing being done. But in general the limitations of human short-term memory are such that we should not be required to remember more than about five unrelated items at one time. How many things does your car require you to remember? Onboard navigation system? Cruise control? Anticollision warning device? Blind-spot warning device? Rearview camera? Entertainment system for the kids? MP3 player? Cell phone? Cars now come with so many of these devices that the systems themselves are contributing to accidents because they increase driver distraction. Yet who gets blamed for the accident—you or the car?

The misattribution of blame is one reason we make the same mistakes over and over again. We learn so little from experience because we often blame the wrong cause. When something goes wrong, especially something big, the natural tendency is to lay blame. But it isn't always easy to figure out where the fault lies. If the mistake is big enough, it will be analyzed by investigators who are presumed to be impartial. But they are plagued by a bias of their own: they know what happened. And knowing *what* happened alters our perception of *why* it happened—often in dramatic ways. Researchers call this effect hindsight bias. With hindsight, things appear obvious after the fact that weren't obvious before the fact.

> **We learn so little from experience because we often blame the wrong cause.**

This is why so many of our mistakes appear—in hindsight—to be so dunderheaded. ("What do you mean you locked yourself out of the house *again*?") It's also why so many of the "fixes" for those mistakes are equally dunderheaded. If our multitasking driver wrecks the car while fiddling with the GPS device on the dashboard, the driver will be blamed for the accident. But if you want to reduce those kinds of accidents, the solution lies not in retooling the driver but in retooling the car.

Much of what we do know about why we make mistakes comes from research in the fields where mistakes cost people their money or their lives: medicine and the military, aviation and Wall Street. When things go wrong, people in these fields have an incentive to find out why. And what they've discovered about their errors can illuminate the reasons for our own. For me, the Aha! moment in this regard came as I worked on a front-page story for the *Wall Street*

*Journal* about the safety record of anesthesiologists. To be sure, the practice of anesthesiology has in recent years benefited from technological innovations. But for a long time, anesthesiologists had a terrible record in the operating room. Their patients often died ghastly deaths. Some suffocated on the operating table because the anesthesiologist, who often became bored during hours-long surgeries, didn't notice the breathing tube had come unhooked. Others inhaled deadly carbon monoxide, an unfortunate by-product formed by the reaction of certain anesthesia drugs. And if those two dangers weren't enough, many of the chemicals used to knock people out were highly explosive. To cut the risk of a spark from static electricity, doctors would wear rubber-soled shoes and put metal grounding pads in their pockets. But every once in a while, Boom! Patient and doctor would be blown to smithereens.

This record, sadly, continued into the early 1980s, when skyrocketing malpractice rates and bad PR (ABC News aired a devastating exposé) forced them to do something. The anesthesiologists, led by a remarkable man, Dr. Ellison C. "Jeep" Pierce Jr., faced a fundamental decision: they could either fix blame or fix problems. They decided on the latter.

Some of the fixes, at least in retrospect, seem obvious. For a long time, there were two major makes of the machines that delivered the anesthesia—essentially Ford and GM, if you want to think of it that way. The models were similar except for one key difference: on the Ford, the valve controlling the anesthesia turned clockwise; on the GM, it turned counterclockwise. Sometimes, anesthesiologists became confused about which model they were working on. They turned the valve the wrong way. The cure was to standardize the machines, so that they all turned the same way.

In other cases, the fixes were more subtle. Anesthesiologists took a page from the pilots' handbook—literally—and began using checklists so they wouldn't forget to do important things. They also engaged in some attitude adjustment. They began discouraging the idea of doctor as know-it-all, and encouraged nurses and others to speak up if they saw someone—especially the anesthesiologist—do something wrong. In error-speak, this is known as "flattening the authority gradient," and it has been shown to be an effective way to reduce errors. In all, these changes required that anesthesiologists acknowledge their own limitations and then do what few of us have the chance to do: they began redesigning their work environments to fit those limitations.

The results have been profound. Over the past two decades, patient deaths due to anesthesia have declined more than forty-fold, to one death per 200,000 to 300,000 cases from one for every 5,000 cases. Their malpractice premiums have also declined—while those of other types of physicians have continued to rise.

"Great," you say. "But unless I need surgery, what does this have to do with me?" Lots, I hope.

### Awareness Is All

As I learned more about the fixes adopted by the anesthesiologists, I began to see connections between their mistakes and ours. Like anesthesiologists, many of us live and work in environments that seem bent, in a thousand small ways, on upping our odds of making a mistake. Just stroll down the aisles of your neighborhood grocery store and notice how the prices are marked. Are the cans of peaches, for instance, marked at twenty-five cents apiece? Or are they four for a dollar? If the price is the latter, you are being subtly manipulated into buying more peaches than you probably need. One study found that when prices are set for multiple units (four for a dollar) instead of single units (one for twenty-five cents), sales increase 32 percent.

As life goes, buying more peaches than you need is not a huge mistake, but it is a telling one. The grocer has, without your knowing it, manipulated the way you analyze a purchasing decision by getting you to anchor your decision on that first number—4. This same effect shapes the way we make not only small decisions, like buying cans of peaches, but, as we'll see, much larger ones, like buying a house.

In the pages that follow, we'll talk about a variety of similar mistakes, from picking the wrong membership plan at your health club to picking the wrong putter at the pro shop. What counts as a mistake? We'll define that term broadly, as the dictionary does:

> **Mistake**—n. 1: a misunderstanding of the meaning or implication of something; 2: a wrong action or statement proceeding from faulty judgment, inadequate knowledge or inattention. **syn:** see error.*

We'll look at why it is, for instance, that you never seem to forget a face—but often can't remember the name to go with it. We'll explore the mistakes made by men and the mistakes made by women (they're often different, as you might guess). And we'll examine some of life's little irritations, like why it is you sometimes have a problem finding a beer in the refrigerator. We'll see how corporations exploit these tendencies, using come-ons like teaser rates on credit cards or offering rebates they know you won't use.

We'll also look at what you can do to make fewer errors. Nothing will make you goof-proof, of course. Many of the tendencies that lead us to make mistakes are so entrenched in our inner workings that it is hard to root them out. It turns out to be very difficult, for instance, to unlearn or ignore bad information—even

---

*Some researchers distinguish between "true" mistakes, which occur when we have the wrong strategy or plan, and more minor errors, known as slips, which occur when there is poor execution of that plan—like stepping on the gas instead of the brake. Still others distinguish between mistakes in process and mistakes in outcome: it is possible, after all, to follow the right process and still get the wrong outcome. But these distinctions, though useful, aren't essential for our purposes.

when we know it is wrong or should be ignored. This has been shown to be true not only in multimillion-dollar negotiations but in everyday decisions, from buying real estate to buying condoms (really).

Nevertheless, there are little things you can do that can make a difference. As with the steps taken by the anesthesiologists, many of these may seem obvious. It helps, for instance, to be well rested—though perhaps not for the reasons you might think. Among other things, sleep-deprived people show a propensity to make reckless gambles (which helps explain why many casinos are open twenty-four hours a day). It also helps to be happy. Happiness fosters well-organized thinking and flexible problem solving, not only in touchy-feely fields like marketing and advertising, but in cerebral ones as well, like medicine. It even helps, believe it or not, to be less optimistic, especially when making decisions. That's because most of us tend to be overconfident, and overconfidence is a leading cause of human error.

Understanding the role of context is also extremely important, especially when it comes to remembering things. Memory, it turns out, is often more a reconstruction than a reproduction. When we try to remember something—a face, a name, a to-do list—it helps if we can be in the same state as when we learned the thing. In one classic experiment, students donned scuba gear and learned a list of words while they were underwater; other students learned words while on dry land. Sure enough, those who learned wet remembered better wet; those who learned dry remembered better dry. The same even held true for those who preferred to drink alcohol; those who learned while slightly intoxicated remembered better if they were tested while tipsy.

> Context matters; those who learned something while intoxicated remembered better if they were tested while tipsy.

Few of us, to be sure, will try to learn anything while underwater, and if history is any guide, even fewer will learn anything while under the influence (though many, I'm sure, will try). But the underlying principles illuminated by these experiments can be applied to our everyday lives in ways that can enrich even the smallest moments. Children, for instance, have been shown to recall vastly more of yesterday's walk in the park if you let them go to the park again (thus reinstating the context) than if you ask them about it in a classroom. Try it with your own kids and see for yourself.

Many of our mistakes are shaped by subtle factors like this. I've come to think of them, collectively, as a kind of trick knee that we all possess: a weakness we can deal with but not eliminate. If we walk a certain way, the knee goes out; if we walk another way, it doesn't, at least not as often. My hope, in discussing these shortcomings in the pages that follow, is that we can learn to walk another way. By gaining a better insight into the things we do well and the things we do poorly, we might do more of the former and less of the latter. Like the Welshmen of St. Brides, we could all benefit from a better understanding of our own limitations.

## Chapter 1

## We Look but Don't Always See

A *man walks into* a bar. The man's name is Burt Reynolds. Yes, that Burt Reynolds. Except this is early in his career, and nobody knows him yet—including a guy at the end of the bar with huge shoulders.

Reynolds sits down two stools away and begins sipping a beer and tomato juice. Out of nowhere, the guy starts harassing a man and a woman seated at a table nearby. Reynolds tells him to watch his language. That's when the guy with the huge shoulders turns on Reynolds. And rather than spoil what happens next, I'll let you hear it from Burt Reynolds himself, who recounted the story years ago in an interview with *Playboy* magazine:

> I remember looking down and planting my right foot on this brass rail for leverage, and then I came around and caught him with a tremendous right to the side of the head. The punch made a ghastly sound and he just *flew* off the stool and landed on his back in the doorway, about 15 feet away. And it was while he was in mid-air that I saw . . . that he had no legs.

Only later, as Reynolds left the bar, did he notice the man's wheelchair, which had been folded up and tucked next to the doorway.

As mistakes go, punching out a guy with no legs is a lulu. But for our purposes the important part of the anatomy in this story is not the legs but the eyes. Even though Reynolds was looking right at the man he hit, he didn't see all that he needed to see. In the field of human error, this kind of mistake is so common that researchers have given it its own nickname: a "looked but didn't see" error. When we look at something—or at someone—we think we see all there is to see. But we don't. We often miss important details, like legs and wheelchairs, and sometimes much larger things, like doors and bridges.

### We See a Fraction of What We Think We See

To understand why we do this, it helps to know something about the eye and how it works. The eye is not a camera. It does not take "pictures" of events. And it does not see everything at once. The part of the visual field that can be seen clearly at any given time is only a fraction of the total. At normal viewing distances, for instance, the area of clear vision is about the size of a quarter. The eye deals with this constraint by constantly darting about, moving and stopping roughly three times a second.

What is seen as the eyes move about depends, in part, on who is doing the seeing. Men, for instance, have been shown to notice different things from those that women do. When viewing a mock purse snatching by a male thief, for instance, women tended to notice the appearance and actions of the woman whose purse was being snatched; men, on the other hand, were more accurate regarding details about the thief. Right-handed people have also been shown to remember the orientation of certain objects

> Women tend to notice details of a woman whose purse is snatched; men notice the thief.

they have seen more accurately than left-handers do. Years ago, after the Hale-Bopp comet made a spectacular appearance in the evening skies, investigators in England asked left- and right-handers if they could remember which way the comet had been facing when they saw it. Right-handers were significantly more likely than lefties to remember that the comet had been facing to the left. Handedness is also the best predictor of a person's directional preference. When people are forced to make a turn at an intersection, right-handers, at least in the United States, prefer turning right, and lefties prefer turning left. As a result, advised the authors of one study, "one should look to the left when searching for the shortest lines of people at stores, banks and the like."

> "One should look to the left when searching for the shortest lines of people at stores, banks and the like."

## The Expert's Quiet Eye

In fact, what we see is, in part, a function not only of *who* we are but of *what* we are. Researchers have demonstrated that different people can view the same scene in different ways. Say you're a golfer, for instance. Even better, say you're a great golfer with a low handicap. You're playing your buddy, who's not so great. You've teed off and played through the fairway, and now it comes time to putt. Do you and your buddy look at the ball in the same way?

Probably not.

Why? Because experts and novices tend to look at things in different ways. One of these differences involves something known as the "quiet-eye period." This is the amount of time needed to accurately program motor responses. It occurs between the last glimpse of our target and the first twitch of our nervous system. Researchers have documented expert-novice differences in quiet-eye periods in a number of sports, ranging from shooting free throws in basketball to shooting rifles in Olympic-style competition. The consistent finding is that experts maintain a longer quiet-eye period.

In the final few seconds of the putt, good golfers with low handicaps tend to gaze at the ball much longer and rarely shift their sight to the club or to any other location. Less-skilled golfers, on the other hand, don't stare at the ball very long and tend to look at their club quite often. Superior vision is so important in golf that many of the world's best players, including Tiger Woods and at least seven other PGA Tour winners, have had Lasik surgery to correct their vision, usually to twenty-fifteen or better. That means they can see clearly at twenty feet what people with twenty-twenty vision could see clearly only at fifteen feet. The sportswear giant Nike has even introduced a new putter, the IC, designed to reduce visual distractions. The shaft and the grip of the $140 putter are both green (to blend in with the color of the grass and reduce distraction), but the leading edge of the blade and the T-shaped alignment line are a blazing white, to help focus a golfer's eyes on the part of the club that contacts the ball.

### We Notice on a Need-to-Know Basis

Regardless of whether we are experts or amateurs, even those of us with otherwise perfect vision are subject to fleeting but nonetheless startling kinds of blindness. One of the most fascinating forms is known as change blindness. It occurs when we fail to detect major changes to the scenes we are viewing during a brief visual disruption—even so brief as a blink.

The profound impact of change blindness was demonstrated a decade ago in an impish experiment by Daniel Simons and Daniel Levin, both of them at the time at Cornell University. The design of their experiment was simple: they had "strangers" on a college campus ask pedestrians for directions. As you might suspect, the experiment involved a twist. As the stranger and the pedestrian talk, the experimenters arranged for them to be rudely interrupted by two men who pass between them while carrying a door. The interruption is brief—lasting just one second. But during that one second, something important happens. One of the men carrying the door trades places with the "stranger." When the door is gone, the pedestrian is confronted with a different person, who continues the conversation as if nothing had happened. Would the pedestrians notice that they were talking to someone new?

In most cases, it turns out, the answer was no.

Only seven of the fifteen pedestrians reported noticing the change.

## Movie Mistakes

At this point, you may find it tempting to think, "I would have noticed a change like that." And maybe you would have. But consider this: you've probably seen countless similar changes and never noticed them. Where? In the movies. Movie scenes, as many people know, are not filmed sequentially; instead, they are shot in a different order from how they appear in the film, usually months or even years apart. This process often results in embarrassing mistakes known in the trade as continuity errors.

Continuity errors have long bedeviled the motion picture industry. The Hollywood epic *Ben-Hur* is a good example. The 1959 movie, which starred the late Charlton Heston as Ben-Hur, won eleven Academy Awards—more than any other movie up to that point in history, including one for Best Picture. But it still has its share of errors, especially in the famous chariot scene, which lasts for eleven minutes but took three months to film. During the chariot race, Messala damages Ben-Hur's chariot with his saw-toothed wheel hubs. But at the end of the race, if you'll look closely, you'll see that Ben-Hur's chariot appears—undamaged! There's also a mix-up in the number of chariots. The race begins with nine chariots. During the race, six crash. That should leave three chariots at the end of the race. Instead, there are four.

Hollywood employs experts who are supposed to catch these things. Officially, they are known as continuity editors or script supervisors, though they are more commonly referred to as script girls because the role, traditionally, has been filled by women. But even they can't catch all the mistakes.

"It's not humanly possible," says Claire Hewitt, who has supervised scripts on a variety of movies, from documentaries and short films to full-length features and even kung-fu action flicks. The best you can do in any given scene, she says, is to try to spot the most important things. But even that is easier said than done.

One of Hewitt's more memorable lapses occurred in her second film as a script supervisor, a short film about a man and a woman who live next door to each other in an apartment building. Instead of filming the actors in separate rooms, though, the filmmakers cheated: they used the same room to film both actors. This required redecorating the room to make it appear in the various scenes to belong to either the man or the woman, but it saved on location costs.

The error occurs in a key scene of the movie, when the woman finally meets the man. "You see her leaning against the door, listening to whether he's out in the hall, and she comes out," says Hewitt. "But the door opens the wrong way!"

Hewitt never noticed the error on her own; it was instead brought to her attention by her mother's boyfriend. "People love doing that—catching you out," says Hewitt. Indeed, entire Web sites are devoted to pointing out continuity errors. (One of the more popular ones is the British Web site moviemistakes.com, run by Jon Sandys, who has been cataloging movie flubs since he was seventeen.) But Hewitt's experience with her mother's boyfriend carries an important lesson: errors that are obvious to others can be invisible to us, no matter how hard we try to spot them.

Okay, you might say, it's easy enough to miss changes to minor details like which way a door opens. Who cares? But what about changes to bigger, more important things?

That's what Levin and Simons wanted to find out. So they shot their own movie. This time, they didn't just change the scenery; they changed the *actors*. During each film, one actor was replaced by another. For example, in one film an actor walked through an empty classroom and began to sit in a chair. The camera then changed, or cut to a closer view, and a different actor completed the action. The films were shown to forty students. Only a third of them noticed the change.

> Only one-third of students shown a short film noticed that the main actor had been changed.

## We See What We Are

When we look at something, we intuitively feel that we can see everything in it in great detail and are quite confident that we would notice any changes. That, said Simons, is what makes change blindness such an interesting problem. "People consistently believe that if something unexpected changes, it will automatically grab their attention and they will notice it." As part of their "door" experiment, for instance, the two Dans polled a group of fifty students. They read them a description of the experiment, then asked them to raise their hands if they believed they would be able to detect the changes. All fifty raised their hands.

The eye, says Simons, has high resolution only at an angle of two degrees. That's not much. If you hold your fist out at arm's length and stick out your thumb, the width of the thumb is roughly two degrees. Superimpose that thumb on a movie theater screen and you get an idea of how little you see clearly. Beyond that, things get progressively blurry. True, you do see some things through your peripheral vision, which is why movies like *March of the Penguins* are popular on wide-screen formats like IMAX theaters. But what you gather

> The eye has high resolution only at an angle of two degrees, or about the width of your thumb held at arm's length; beyond that, things get blurry.

through this peripheral vision, says Simons, is broad, blurry information. "You're not going to see the details of the penguins."

The details we do notice depend, to a degree, on how we define ourselves. In the door experiment, for instance, Simons and Levin found that the seven pedestrians who did notice the change had something in common: they were all students of roughly the same age as the "stranger" they encountered. In one sense, this finding wasn't surprising. Social psychologists have shown that we often treat members of our own social group differently from how we treat members of other groups. Black people encountering white people (or vice versa) may behave differently than when they encounter someone of their own group; ditto for rich people encountering poor people, old versus young, and men versus women. Nonetheless, wondered Simons and Levin, would those differences in the way we *behave* toward others extend to the way we *see* others?

To answer that question, they repeated the door experiment, using the same "strangers" they had initially used. Only this time the strangers weren't dressed casually, as students would be; they were dressed as construction workers, complete with hard hats. And this time, they approached only people of their own age. In all, the "construction workers" encountered twelve pedestrians. Of those twelve, only four reported noticing the switch when the door came through. Putting the experimenters in construction clothes, it seemed, had been enough to change the way they were seen by students. Rather than being seen as individuals—as they had been when they were dressed as students—the experimenters were now seen as members of another group.

One of the pedestrians who had failed to detect the change when the door was brought through said as much when she was told of the experiment and interviewed afterward. She said she had seen only a "construction worker" and had not really noticed the individual; that is, she had quickly categorized him as a construction worker and hadn't noted those details— like his hair or his eyes or his smile—that would allow her to see him as an individual. Instead, she had formed a representation of the category—a stereotype. In the process, she traded the visual details of the scene for a more abstract understanding of its meaning; she had skimmed.

> We trade visual details for a more abstract understanding of meaning. In other words, we skim.

As we'll see later, we skim quite a lot. And for most of us, skimming works just fine, at least most of the time. If we pass a construction worker on the street, we probably don't need to study his face. After all, we don't need to know *who* he is; we just need to know *what* he is. If we were aware of making such a distinction at the time we were making it, that would be one thing. But the problem is, we think we've noticed when we haven't. We don't know when we're skimming.

## You Can't Handle the Truth

Knowing that we are susceptible to visual errors like change blindness does not allow us to compensate for their effect; we're still vulnerable. I can't demonstrate this by running a door by you, but let's

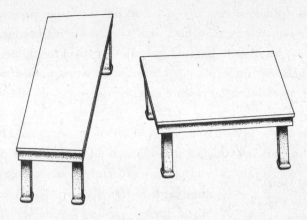

try something else. Take a look at the tabletops above. Which one is bigger?

The answer is neither. As hard as it is to believe, the tabletop on the left is identical in size and shape to the one on the right. You can check by cutting out a piece of paper the exact shape of one and placing it over the other.*

The interesting thing is that knowing this is an illusion doesn't allow us to correct for its effect. No matter how many times we look at the tables, they *still* appear to be shaped differently.

The tabletops are the work of the renowned Stanford professor Roger N. Shepard. A prankster from a young age (he once secretly moved all the furniture out of his sister's room), Shepard has long delighted in using visual tricks like this one to make larger, more serious points. In this case, his illustration, called "Turning the Tables," is used to demonstrate not only that our perceptual machinery is deeply entrenched in our nervous system but that its operation is wholly automatic. As a result, we cannot choose to see a drawing merely as what it is—a pattern of lines on a flat piece of paper. Instead, under the right conditions, the pattern automati-

*Or, for an animated demonstration, go the Web site of Michael Bach: www.michaelbach.de/ot/sze_shepardTables/index.html.

cally triggers circuits in the brain that make a three-dimensional interpretation. What's more, this effect passes unnoticed. Until one is told that the two seemingly differently shaped tabletops are identical, there is no reason to suspect anything is amiss; we've made an error, but we don't *know* we've made an error.

## We See What We Expect to See: The Case of the Missing Beer

A final point is worth noting: what we see also depends, in part, on what is looked for. By and large, we see what we expect to see. Commonly seen items get noticed a lot; rarely seen ones don't.

"If you don't find it often, you often don't find it," says Jeremy M. Wolfe, a professor of ophthalmology at the Harvard Medical School. Dr. Wolfe's specialty is the field of visual search. Researchers in this field try to answer what has been described as the beer-in-the-refrigerator problem: How do we find the things we are looking for?

> "If you don't find it often, you often don't find it."

The answer is not as straightforward as it might appear. You might search for the beer by looking on a certain shelf of your refrigerator, because you know the beer is usually placed on that shelf. But what if the beer gets moved to make room for other groceries? In that case, you might search by the shape of the bottle or can. But other items in the refrigerator might have the same shape; a can of Coke, for instance, might resemble a can of Budweiser. You might hunt long and hard before you find what you are looking for.

## We're Built to Quit

Seeing, it turns out, is very hard work. Just how hard is difficult for most of us to appreciate. For those of us who have always been able to see, nothing seems more natural: you simply open your eyelids and, presto, there is the world. But for those who were once blind, learning to see can be a harrowing experience. Before the outbreak of World War II, the German researcher Marius von Senden collected and published accounts of nearly a hundred individuals across the Western world who had been blinded by cataracts that were later surgically removed.

For many of the patients, learning to see proved to be a wrenching experience. One man, upon venturing out to the streets of London, "confused his sight to such an extent that he could no longer see anything," reports von Senden. Another man could not judge distance. "Thus he takes off one of his boots, throws it some way off in front of him, and then attempts to gauge the distance at which it lies; he takes a few steps towards the boot and tries to grasp it; on failing to reach it, he moves on a step or two and gropes for the boot until he finally gets hold of it." One boy found learning to see so difficult that he threatened to claw out his eyes. Discouraged and depressed, many others simply quit trying to see.

Something similar happens to people when they look for things that are seldom seen. In a recent experiment Dr. Wolfe and his colleagues at the Visual Attention Lab at Boston's Brigham and Women's Hospital asked volunteers to look at thousands of images. Each image was set against a busy background filled with other images (the refrigerator). The volunteers were then asked to report whether they saw a tool, like a wrench or a hammer (the beer).

When the tool was present a lot—which was true half the time—the volunteers did a great job of spotting it. They were wrong only 7 percent of the time. But when the tool was rarely present—say, in only one out of every hundred images—their performance went way downhill. Their error rate soared to 30 percent.

Why? They gave up. Wolfe demonstrated that observers have a quitting threshold—basically, the amount of time they will look for something before giving up. Typically, observers slow down after making mistakes and speed up after successes. Since observers looking for seldom-seen items can successfully say no almost all the time and still be right, they tend to speed up and drive down their

> **Observers have a quitting threshold—basically, the amount of time they will look for something before giving up.**

quitting time. Before long, they're like Fred Flintstone leaving the quarry: yabba-dabba do.

In fact, Dr. Wolfe found, the people in his experiments abandoned their search in less than the average time required to find the target. Like the formerly blind people studied by von Senden, they simply quit trying to see.

In any case, says Wolfe, the moral of this story is that we are built—perhaps hardwired—to quit early when the target is unlikely to be there. And most of the time, he says, that works well enough. "I mean, it's really dumb to spend vast amounts of time searching for things that aren't there."

## Please Stow Your Gun in the Overhead Compartment

Unless, of course, it's your job to spend vast amounts of time looking for things that usually aren't there. For instance, what if your job were to find a gun? Or a tumor? People don't want you to quit early—they want you to stay late.

Both baggage screeners at airports and radiologists at hospitals spend the bulk of their time looking for things they rarely see. In the case of radiologists, routine mammograms reveal tumors only 0.3 percent of the time. In other words, 99.7 percent of the time, they won't find what they're looking for. Guns are even rarer. In 2004, according to the Transportation Security Administration, 650 million passengers traveled in the United States by air. But screeners found only 598 firearms. That's roughly one gun for every million passengers—literally, one-in-a-million odds.

Both occupations, not surprisingly, have considerable error rates. Several studies suggest the "miss" rate for radiologists hovers in the 30 percent range. Depending on the type of cancer involved, though, the error rate can be much higher. In one especially frightening study, doctors at the Mayo Clinic went back and checked the previous "normal" chest X-rays of patients who subsequently developed lung cancer. What they found was horrifying: up to 90 percent of the tumors were visible in the previous X-rays. Not only that, the researchers noted, the cancers were visible "for months or even years." The radiologists had simply missed them.

> In one study, radiologists missed up to 90 percent of cancerous tumors that, in retrospect, had been visible "for months or even years."

As for the nation's fifty thousand airport screeners, the federal government won't reveal how often they make mistakes. But a test in 2002 indicated that they missed about one in four guns. During a similar test two years later at Newark's airport, the failure rate was nearly identical: 25 percent. More recently, 60 percent of bomb materials and explosives hidden in carry-on items by undercover agents from the TSA were missed in 2006 by screeners at Chicago's O'Hare International Airport. At Los Angeles International Airport, the results were even worse: screeners missed 75 percent of bomb materials.*

And keep in mind, these are trained professionals dealing with life-or-death issues. What about you or me? How good would we be at picking out important things in the world around us? Like the face of someone who attacked us?

---

*A spokeswoman for the TSA said performance of screeners has improved since 2006, in part because the TSA had employed a wider variety of tests. But she declined to give a more current miss rate because "the results of these tests are not appropriate for public dissemination" (*Chicago Sun-Times*, Oct. 19, 2007).

*Chapter 2*

## We All Search for Meaning

$I$*n the 1970s*, the noted psychologist Harry Bahrick conducted a landmark study that will interest anyone who has recently attended a class reunion—or plans to. Bahrick and his colleagues asked hundreds of former high school students to look back at their yearbooks and see whether they could remember the faces of their classmates. What they discovered is a tribute to the power of human memory. For decades after graduation the memory of former students for the faces of their classmates was nearly unimpaired. Even after nearly half a century had elapsed, the former students could still recognize 73 percent of faces of their classmates.

But when it came to names, Bahrick found, memories were much worse; after nearly fifty years the former students could remember only 18 percent of their classmates' names. Names, for whatever reason, do not stick very well in our memories, or they stick only partway, causing us to call our brother-in-law Bob, Rob, or to mistake the author Ernest Hemingway for the actor Ernest Borgnine.

### Meaning Matters; Details Don't

Why should we remember faces, but not the names that go with them? Part of the answer is that when it comes to memory, meaning is king. Our long-term memory, even for things we've seen thousands of times, is limited. It is primarily semantic, which means that in most daily instances of remembering what we must recall is meaning, not surface details. Take the common penny, for instance. How well do you think you can remember its features? In a well-known test, two researchers, Raymond Nickerson and Marilyn Adams, asked just such a question. The answer they got surprised them—and may surprise you.

In the test, Nickerson and Adams asked twenty people to do something that sounds deceptively easy: from memory, draw the front and back of a penny. (If you like, take a few minutes yourself to do this before reading what comes next. Don't cheat by looking at a penny first.) After the drawings were done, Nickerson and Adams graded them to determine how accurately the participants had depicted eight critical features, like the placement of Lincoln's profile on the front of the coin and the placement of the Lincoln Memorial on the back.

The results were lousy.

Of the twenty people tested, only one—an avid penny collector—accurately recalled and located all eight features (presumably, these features held special meaning to him). Of the eight features, the median number recalled and located correctly was just three. Interestingly, the most frequently omitted feature was the word "LIBERTY," which appears on the front of the coin, to the left of Lincoln's profile. (If you've drawn your own penny, take out a real one now and compare the two. Did you get more than three features right?)

> Of twenty people tested, only one—an avid penny collector—could accurately recall and locate eight critical features of a penny.

The findings from the penny-drawing test were surprising enough that Nickerson and Adams conducted a series of follow-up tests to try to pin down what was going on here. Among other things, they wondered: If people couldn't recall exactly what a penny looks like, would they at least be able to tell the real thing from a fake?

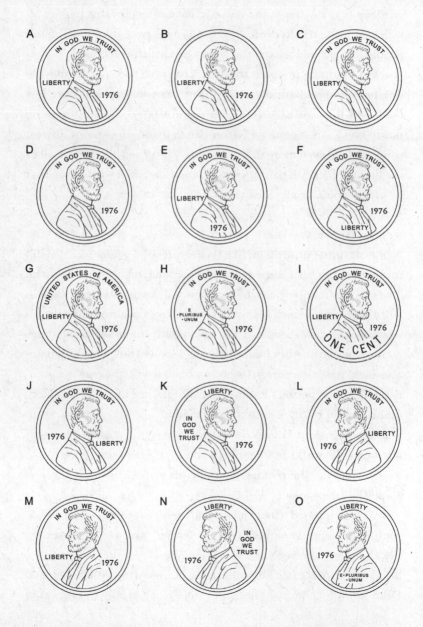

To find out, they showed a new group of people fifteen drawings of the heads side of a penny. Only one of the drawings was accurate; the rest were not. The participants' job was to pick the right one. See if you can spot the real McCoy on the previous page.

Again, the results were disappointing. Fewer than half of the people in the study picked the right one (which is A).

You might be tempted to think that these poor results may be due to some peculiarly American trait (maybe Americans are less observant than members of other cultures are?). But this appears not to be so. A subsequent test in Britain, using images of British coins, yielded even more dismal results. "It turned out that recall of the design of British pennies was, if anything, even worse than that of U.S. pennies," concluded the study's author.

### Names Don't Matter

Names, it turns out, are a lot like the features of a penny—they don't mean much, and as a consequence we tend to forget or confuse them. The relative unimportance of names was demonstrated years ago in a British study in which researchers asked real people to study the biographies of fake people. Each biography contained a fictitious name, along with other phony information, like the name of a place associated with that person (such as a hometown), the person's occupation, and the person's hobby. For example, one of the bogus biographies might read, "A well-known amateur photographer, Ann Collins, lives near Bristol, where she works as a health visitor."

So what did the real people remember about the fake people?

If you guessed their jobs, you were right. Jobs were remembered 69 percent of the time. In close second were their hobbies, with 68 percent. Then came their hometowns, with 62 percent. Dead last were names. First names were recalled only 31 percent of

> It is easier to remember that someone is a baker than it is to remember that his last name is Baker.

the time and last names 30 percent. Other studies have found similar results. For some reason, it is easier to remember that someone is a baker than it is to remember that his last name is Baker.

Why should this be? Researchers aren't sure. But the best guess is that names, in and of themselves, don't mean much; they're just arbitrary labels. Jim or Tim, Anne or Fran—there's no inherent meaning in any of these names, at least not for most of us. Jobs, hobbies, and places, on the other hand, are often "semantically richer"—they *mean* something. Maybe you've been to Bristol, for instance, or fancy yourself a photographer. If so, those qualities will tend to stick in your memory; they have meaning. But names don't.

## A Slip of the Tongue?

This small insight reveals much about a mistake that afflicts us all: failing to come up with the name of a person we know or—even more embarrassing—calling someone by the wrong name. Researchers call these gaffes slip-of-the-tongue or tip-of-the-tongue errors, or TOTs for short, and they are exceedingly common. They have been documented across all ages and cultures, and for most people they occur about once a week.

One of the more infamous slips, at least in recent times, occurred a few days before the 1992 Super Bowl. Joe Theismann, a former quarterback for the Washington Redskins, found himself being interviewed by two newspaper reporters about the coach of the Redskins, the legendary Joe Gibbs. Gibbs was—and still is—considered one of the finest offensive strategists ever to have stepped onto a football field. And the reporters wanted to know whether Theismann thought Gibbs was some kind of genius.

Theismann didn't think so. In the first place, he told them, the word "genius" isn't applicable in a sport like football.

"A genius," said Theismann, "is a guy like Norman Einstein."

*Norman* Einstein? Theismann clearly meant to say *Albert* Einstein. But it was too late. His slip, which was picked up by *Sports*

*Illustrated*, made national news. Before long, Theismann became the poster child for dumb jocks everywhere. But his remark, as we'll see, wasn't as dumb as it first appeared to be.

Research on TOTs shows that most of these mistakes involve proper names—like Albert Einstein. Why proper names? The answer isn't certain. But the best guess is this: when your brain is searching for a proper name, only one name will do. If you are trying to think of the capital of North Dakota, for instance, only one word will get you there: "Bismarck." But if you are searching for a common noun, synonyms are available. If you are trying to think of the term for the part of a computer that displays text, for example, and you can't remember the word "monitor," you can substitute the word "screen."

## We Recall Part, but Not All

The intriguing part of this is that often when a proper name is on the tip of our tongues, we can recall *some* of the information we need. Moreover, this partial information is of a particular type. People who can't quite come up with the right name, for instance, will often be able to guess the right number of syllables in the name and even the first letter of the name. In one study, for instance, a person tried to identify a picture of the actress Liza Minnelli. The person couldn't quite come up with her name, but tried to get at it by writing out names that came tantalizingly close:

1. Monetti
2. Mona
3. Magetti
4. Spaghetti
5. Bogette

Another clue to the TOTs riddle is that recall of the right name is often blocked by a wrong name that is nevertheless persistently brought to mind. But it's not just any wrong name. The wrong name typically has the same *meaning* as the right name; if you're thinking about a smart person like Albert Einstein, for instance, the wrong name will likely be that of another person you also consider very smart. And this is where the Theismann story gets interesting.

It turns out that there really is a Norman Einstein. He is an emergency room physician at Catawba Valley Medical Center in Hickory, North Carolina, where, Dr. Einstein told me, "no one has ever mistaken me for a genius."

He and Theismann, as it happens, were high school classmates; both attended South River High School in New Jersey, though two years apart.

"I was a senior when he was a sophomore," Einstein said. As boys, the two men lived maybe five or six blocks apart. "We played a little bit of basketball, touch football—that kind of stuff," Einstein said, but they weren't close friends. Theismann was a jock, and Einstein was a brain. Einstein graduated in 1965 and was the class valedictorian. He attended Rutgers University, where he majored in physics, and then went on to medical school at Tufts University. Theismann, meanwhile, headed to the NFL. And so their paths diverged until, twenty-seven years later, in a corner of the Minneapolis Metrodome, Norman Einstein's name popped back into Joe Theismann's mind. And in Theismann's mind, the surface details regarding Norman Einstein and Albert Einstein may have faded, but their common meaning had not: both were two very smart guys. And what Theismann had recalled was that meaning. In fact, after *Sports Illustrated* learned that there actually was a Norman Einstein and that he was a very smart guy, Theismann told the magazine, "My comment was not as absurd as it might have seemed."

## We Can Make the Meaningless Meaningful

Try as we might, it's extremely difficult to force our minds to remember meaningless things. This difficulty was quantified over a century ago by a young German named Hermann Ebbinghaus. Ebbinghaus spent years of his life memorizing thousands of nonsense syllables like DAX and QEH. Morning, noon, and night, year after year, he would rehearse long lists over and over again to the beat of a metronome—MEB, FUT, PON, DAK, GOL, LIG—until he had them memorized. Despite the headaches and exhaustion that would often follow such sessions, Ebbinghaus would wait for an interval of time to pass, then test his memory. He found that when the syllables made no sense—as was the case with his experiments—they were quickly forgotten. After just one hour, for instance, Ebbinghaus had forgotten more than half of the syllables that he had, with great effort, attempted to learn.

To a degree, we can overcome such forgetting by reframing otherwise meaningless information in a way that imbues it with meaning. One study, for instance, tracked the remarkable skill of a long-distance runner—not for running, but for remembering. The runner, who was at the time a young college student, had no special gift for memory; he was of average intelligence and scored a middling 990 on his Scholastic Aptitude Test. But through years of often daily practice, he developed an extraordinary ability to remember strings of numbers. Over the course of two years he was able to increase his digit span from seven digits, which is about what most people can remember, to eighty. Not only is that more than ten times the normal digit span—it was four times higher than had ever been recorded.

How'd he do it?

Researchers found that whenever possible, the runner had memorized sequences of three digits not as separate numbers but as running times. For instance, the numbers 5-1-3 would be memorized as 5 minutes, 13 seconds—a good time for a mile-long race. In other words, the runner had converted previously meaningless information into information that had meaning, at least to him.

This is an old trick, known as mnemonics, and it has been around since at least the days of the Greeks. The Greeks, like every civilization, needed a way to convey information from person to person. Since the printing press hadn't been invented, there were no books. That left them with that old standby, word of mouth. But for word of mouth to be accurate, people needed a way of memorizing large chunks of information, sometimes very large. So the Greeks learned to associate the meaning*less* with the meaning*ful*.

You can do something similar, by the way, and amaze your friends in the process. Try to memorize the following string of twelve meaningless numbers: 1, 7, 7, 6, 1, 8, 6, 5, 1, 9, 4, 5. Not too easy, right? Now break them up into three meaningful dates in American history: 1776, 1865, and 1945. Much easier, no?

But this trick has its limits, as the long-distance runner discovered. When he was presented with sequences of three numbers that could *not* be memorized as meaningful running times—say, 4-8-3 (which would be 4 minutes, 83 seconds, a nonsense time)—his digit span plummeted, and he quickly forgot what he was supposed to remember.

## Why We Forget Passwords and Hiding Places

We run the same risk when we try to outfox ourselves by, say, hiding valuables in a clever spot or picking passwords that nobody will ever guess. If the hiding spot or password lacks meaning, we will soon forget it, just as Ebbinghaus did, no matter how hard we try to remember it. We see examples of this every day. For instance, the *New York Times*, whose customers include some of the most well-educated readers in the world, reported that one thousand online readers *each week* forget their passwords. In addition, up to 15 percent of its "new" users were actually old users signing up again because of a forgotten password. And the *Times*'s experience is in no way unusual. By one estimate, up to 80 percent of all calls to corpo-

> A recent poll of three thousand people found that one-fourth of them couldn't remember their own home phone numbers.

rate computer help desks are for forgotten passwords.

Passwords aren't unique. Our lives are filled with other important things that we nevertheless forget with alarming frequency—like birthdays and anniversaries, wallets and cell phones, and the spot where we parked the car. A recent poll of three thousand people found that one-fourth of them couldn't remember their own phone numbers, and two-thirds couldn't recall the birthdays of more than three friends or family members.

Yet, as overloaded as we are with things to remember, we often persist in picking hiding places we are doomed to forget. In one survey, more than four hundred adults were asked whether they had recently found an object that they had lost or misplaced. Of those who had recalled such a recent episode, 38 percent reported finding the item in a place that was not "logical." Why would such a high percentage of lost items be found in illogical places? Researchers concluded that people mistakenly believe that the more unusual a hiding place is, the more memorable it will be. But the opposite turns out to be true: unusualness doesn't make a hiding place more memorable—it makes it more *forgettable*.

Tom Vander Molen found this out the hard way. When Tom, who lives in Grand Rapids, Michigan, was five years old, his grandparents gave him a gold coin. That was back in 1963, and his grandparents, who were not wealthy people, called Tom and his older brother into the living room of their Cincinnati home.

"We want to give you something special," they said, and handed the boys a small cardboard box stuffed with cotton. Inside were two $5 gold pieces—one for Tom and one for his brother. Tom had never seen gold before.

"I was in awe," he said.

Within a few months, Tom's grandfather died. A few years later, Tom's father also passed away. Then, in Tom's senior year of high school, doctors discovered a tumor at the base of his spine. Soon, he was restricted to life in a wheelchair. Through those lonely and painful days, one of the things that sustained Tom was that gold coin from his grandparents and the fond memories it gave him.

Over the years Tom began to buy more coins and add other valuables to his collection. By 1995, Tom figured, his stash was worth maybe $4,000.

But now he faced a difficult decision: Where to hide it?

The key to picking a good hiding place is making a quick connection between the thing being hidden and the place in which it is hidden, says Alan Brown. Brown is a professor at Southern Methodist University who has studied where and how people hide things. Not long ago he surveyed adults between the ages of eighteen and eighty-five, asking them all sorts of questions about where they hide things. Their answers have provided some illuminating differences. Older adults, for instance, typically hide jewelry from thieves, whereas younger adults tend to hide money from friends and relatives. And while the places they choose may vary, the successful strategies didn't.

"I think the only successful way to do it—and this is true with both hiding places and passwords—you have to do it quickly," said Brown. "You don't have ten or twenty minutes to figure it out scientifically. You have to come up with it on the spot."

> One key to picking a good hiding place— or a good password: do it quickly.

But this is not what Tom Vander Molen did.

"I just sat there for a while and thought, 'What's a good, safe place? Where will a break-in burglar not look?'"

That's when he noticed the paint cans. There, in the bottom

of a small metal cabinet in his storage room, were a bunch of gallon paint cans. Many were half-empty and filled with dried-up paint.

"Aha!" he thought. No burglar would ever think to look there. So he pried open one of the cans of dried-up paint and dropped his stash inside. It was, he remembers thinking, "the perfect hiding place"—so perfect that he never gave it a second thought. Over the next eight or nine years, Tom never went back to check on his stash.

"Not once," he said.

But then, in the spring of 2004, he found himself wondering, "My coins? Where are my coins?" But for the life of him, he could not remember. After a while, he could stand the wondering no more.

"I said, 'I'm going to stay up as late as I have to. I'll look under every single box, under every shelf unit, through the dust bunnies, in the back of my fridge—I'm going to look everywhere.' "

And he did. He even looked *in* the fridge. "I took all of the frozen stuff out of the freezer, and I looked through every bit of wrapped stuff to say, 'Okay, was it in here?' I just kept going, turning up nothing."

Then, about one in the morning, Tom wheeled himself into the storage room, opened the door to the metal cabinet, and said, "Aha! I've found it."

But just as quickly, his epiphany vanished.

That summer a friend had stopped by to help Tom do some painting. One of the first orders of business had been to clear out the old cans of dried-up paint.

"Let's just put them all together," Tom said to his friend, "and I'll take them to work and chuck them in the Dumpster."

And that's exactly what he did. Without a moment's hesitation, Tom heaved the paint cans into the Dumpster. A few days later the contents of the Dumpster were emptied, then hauled away to be crushed and incinerated; among them were Tom's gold coins.

"It was like the perfect loss," Tom told me, "because I was both the victim and the perpetrator."

## Making Faces More Memorable

That's often the way it is with human error—we end up being both the victim and the perpetrator. But it doesn't have to be that way. The case of the long-distance runner that I mentioned a few pages back shows how, with great practice, we can imbue otherwise meaningless information with meaning and thus make something more memorable. Sometimes, though, our brains do this for us, automatically extracting meaning from the world even when we have no idea they are doing so.

This is especially true when it comes to people's faces. Humans seem to have an innate knowledge of faces. Experiments with newborns, for instance, have shown that almost from the moment we are born, the faces of other people hold a special attraction for us. As adults, we can often make uncannily accurate judgments about people after looking at their faces for only a fraction of a second. Moreover, our ability to recognize others does not rely on their physical features alone—though we often think it does. Most people, when asked how it is they recognize someone, will almost invariably identify some physical feature. When you are trying to figure out whether you recognize someone, for instance, what do you look for? Many studies have sought to answer this question. The most consistent finding among them is that the single most important feature is . . . hair. Which is an interesting choice given that hair, of all our physical features, is the one most easily altered; it can be cut, dyed, grown, and even, alas, lost. But hair it is.

When researchers asked a slightly different question, though, they got a surprising answer. They found that when faces are

> If you want to remember someone, try judging his face for emotional traits, like honesty.

judged not by their surface details but for deeper emotional traits—like honesty or likability—the faces are subsequently better recognized than faces judged for physical features like hair or eyes. Why should traits be more memorable than features? Traits appear to require the brain to engage in a greater depth of processing; it takes more work to figure out whether someone has an honest face than it does to determine, say, whether he's got curly hair. And that greater effort seems to make the face stick in the memory. So strong is the effect that one of the leading researchers on face recognition once offered this advice: "If you want to remember a person's face, try to make a number of difficult personal judgments about his face when you are first meeting him."

## How Not to Identify a Suspect

June Siler got just such a chance on the evening of February 28, 1997. Only twenty-four and newly arrived from a small town in Michigan, Siler had just finished a twelve-hour shift as a registered nurse at Michael Reese Hospital, on Chicago's South Side. It was Friday—a payday Friday—and Siler usually treated herself to a cab ride home on paydays. But this Friday was different. She was going on vacation and thought she would save the $15 in cab fare. So Siler, still in her white nurse's scrubs, put on her jacket, shouldered a small backpack, and walked the few blocks to the bus stop. As she leaned against a pole of the bus shelter, a man walked past her and stood a few feet away.

"How long you been here?" he asked.

"Just arrived," she said.

A few minutes passed; neither person spoke. But Siler kept her eye on the man. Michael Reese is in a tough neighborhood, and the man next to her bore watching. She particularly noticed his footwear: black Reebok athletic shoes covered by Velcro straps. She felt sorry for him; the shoes were so out of style, she thought, that he must have gotten them from a homeless shelter.

Siler turned her head to look for the bus. When she turned back, the man grabbed her. He clamped her in a headlock and began slashing her across the neck. He kept slashing and slashing— across her face, across her neck, across her chest. It happened so quickly she wasn't sure what was going on. But then, as the blade cut across her eye, she heard the oddly familiar clink of a box cutter. That's when she knew she was in trouble.

Siler tumbled backward into the street, toppling the man and knocking loose the box cutter. Then she stood up and, for the first time, looked him hard in the face.

"Fuck you!" she screamed. "Fuck you!"

Just then, the light changed.

Traffic started to rise up toward them, headlights illuminating the fray. The man fled, leaving Siler alone in the street. During the attack her backpack had been torn off and its contents spilled. So she gathered her mittens and her newspaper, put them back into the backpack, and began to walk to the hospital's emergency room.

"I thought it was starting to rain because I could hear drops hitting," she said. But it wasn't rain; it was her blood.

As she lay in a hospital room, Siler gave police a detailed description of the man who attacked her. Within twenty-four hours, detectives stopped a man standing next to the same bus shelter where Siler had been attacked. His name was Robert Wilson. He not only matched the description Siler had given, but when the officers frisked him, they found a pistol—and a knife. He was booked and photographed, and when police showed the photographs to Siler, she identified him as the man who had attacked her.

When the case came to trial, the prosecutor asked Siler to once again identify the man who attacked her. But this time she balked. She was no longer looking at a photograph, but at a flesh-and-blood human being. And as she stared at Wilson from the witness stand, Siler searched not so much for the physical features of the man— not his nose, or his eyes, or his hair; what she looked for was the emotional trait she had encountered the night of her attack.

"I wanted to see the hate that I saw at the bus stop," she said. "I wanted to have the feelings that I felt at the bus stop."

But no matter how hard she looked for those traits, she could not find them.

"I had nothing," she said.

That nothingness should have tipped her off that Robert Wilson was the wrong man. But the police and prosecutors were confident that Wilson was the right man; he had even signed a confession. And so, when the prosecutor asked her to identify the man who attacked her, Siler stood and pointed to Wilson. A short while later, he was convicted of attempted murder and sentenced to prison for the maximum term: thirty years. As the trial ended, the judge made special note of Siler's testimony. She was, he said, "the most solid, positive, outstanding victim witness I have ever seen."

After the trial, Siler left Chicago. But one fact continued to eat at her: the police had never been able to find those black Velcro gym shoes that she had seen her attacker wearing that night at the bus stop. Even though police had searched Wilson's apartment, they had not come up with them. And she felt sure that the man who attacked her wouldn't throw them away; he would have been too poor to afford another pair.

Then, one day in 2006, Siler got a phone call from a newspaper reporter. He told her that Robert Wilson had appealed his conviction, and won. The judge ruled that the trial court had improperly excluded the evidence that another man—who strongly resembled Wilson—could have attacked Siler. Most important of all, the judge noted, at the time of his arrest the other man was wearing black Velcro gym shoes—"exactly like the ones described by Siler."

As soon as Siler heard about the shoes, she said, "I knew."

She broke down and began to sob. After hanging up with the reporter, Siler called Wilson's attorneys and offered to do whatever she could to get him out of jail. A few weeks later, Robert Wilson was freed.

## The Ugly Face of Crime

The story of Robert Wilson and June Siler is, of course, part of a much larger story—the persistent failure of eyewitness identification. Recent studies suggest that misidentifying one person for another is more common than many of us would like to believe. Between 1989 and 2007, for instance, 201 prisoners in the United States were freed through the use of DNA evidence. Of these, 77 percent had been mistakenly identified by eyewitnesses.

Why should this be? Why should a stranger be mistaken for someone we know—or think we know? One clue seems to lie in the meaning we take away from such encounters. When we make personal judgments about someone, these judgments form relatively strong bonds in our memory; they stick. Yet when we are called upon to identify someone, we often ignore these traits in favor of more traditional features that are less strongly bonded in our memories—like hair or eyes or race. As we saw in the first chapter, we often do not recognize "others" as intimately as we do members of our own group. Studies have shown that the faces of people belonging to our own race, for example, are more easily recognized than faces of people belonging to other races; they "mean" more. So does beauty. Pretty faces, it seems, are more easily recognized than ugly ones. In one study, three hundred people were divided into groups. The groups consisted of old and young, black and white, male and female. Each was asked to look at pictures of students and faculty taken from school yearbooks and see which faces they recognized. "In every case," the authors concluded, "a higher percentage of subjects who perceived beauty in a face were subsequently able to recognize the face."

This may help explain why it can be difficult to identify criminals. When we speak of "the ugly face of crime," it is not entirely a metaphor: recent research indicates that criminals are, by and large, uglier than the rest of us. Two professors,

> Recent research indicates that criminals are, by and large, uglier than the rest of us.

Naci Mocan of the Louisiana State University and Erdal Tekin of Georgia State University, analyzed data from a federally sponsored survey of fifteen thousand high schoolers who were interviewed in 1994 and again in 1996 and 2002. One question asked interviewers to rate the physical appearance of the student on a five-point scale ranging from "very attractive" to "very unattractive." The professors found that the long-term consequences of being young and ugly were small but consistent. "Unattractive individuals commit more crime in comparison to average-looking ones," they concluded, "and very attractive individuals commit less crime in comparison to those who are average-looking."

Yet pretty faces are the ones we tend to recognize.

*Chapter 3*

# We Connect the Dots

One *way to think* about June Siler's experience is in terms of Roger Shepard's illustration. Shepard, you'll remember, is the man who created the illustration we saw in the first chapter, "Turning the Tables." Shepard believed that the machinery that allows us to perceive the world around us is not only deeply embedded inside us; it's also automatic. That's why we can't choose to see a drawing like his merely as what it is—a pattern of lines on a flat piece of paper. Instead, the brain turns an object drawn in two dimensions into one that appears to have been drawn in three. In effect, the brain connects the dots we didn't know it was connecting.

Something similar appears to happen in that moment when we experience a flicker of recognition. When Siler was on the stand, her brain was trying to make the association between the hate she experienced on the night she was attacked and the traits she perceived that day on the stand. And the two didn't add up. In short, she was trying to connect dots that that deeply embedded machinery couldn't connect. And the machinery was right: she had the wrong man. But Siler tried to override this impulse, and when she did, she made a mistake.

## Snap Judgments Are Hard to Shake

These types of subtle connections are more powerful—and common—than many of us would like to believe. Take a look at the two faces above, for instance.

Which person is more competent?

If you're like most people, you picked the man on the left. That's whom voters picked, too. His name is Russ Feingold, and he is a U.S. senator from Wisconsin. In 2004, Feingold defeated the man on the right, the Republican Tim Michels, by a wide margin, 55 percent to 44 percent.

Feingold's picture was part of a study by Alex Todorov and other researchers at Princeton University's Woodrow Wilson School of Public and International Affairs. They showed black-and-white photographs of political candidates to a number of participants. (If any of the participants recognized the faces, their opinions were eliminated from consideration in the study.) They found that faces are a major source of information about other people. In particular, they showed that in-

> Inferences about the competence of politicians occurred within one second of being exposed to their faces.

ferences of competence based solely on facial appearance predicted the outcomes of U.S. congressional elections better than chance. The candidate who was perceived as more competent won 72 percent of the Senate races and 67 percent of House races.

More important, at least for our purposes, people drew these inferences very quickly. In a subsequent test, the inferences about the competence of politicians occurred within *one second* of being exposed to their faces. Not only that, the researchers found that these inferences didn't change much if people were given more time to think about it; their initial impression stuck.

It's not clear what's behind these impressions: Is it the jut of the jaw or the set of the eyes that makes someone look more competent? We don't know. But there seems to be something to the impressions we form about people's faces. One study of cadets at West Point, for instance, found that judgments of facial dominance, as measured from the cadets' graduation portraits, predicted military rank attainment; those with sterner faces eventually outranked those who were less formidable looking.*

### A Woman's Body Gives Her Away

Often, the cues we use to form judgments about other people are so subtle as to be all but invisible. One example comes to us from an unlikely source of such information: topless dancers. Their income, it turns out, depends not only on the attributes we might expect but on something far less obvious: their fertility cycle.

> The income of topless dancers depends not only on the attributes we might expect but on something less obvious: their fertility cycle.

*This is not to say that these impressions are always accurate. In another study, baby-faced people who were judged as being less competent than people with more mature faces actually tended to be more intelligent. See Montepare and Zebrowitz (1998).

It is well-known that women undergo physical changes over the course of their menstrual cycle. Body odor changes, as does facial attractiveness and body shape. Near the most fertile point of a woman's cycle (just before ovulation), for instance, lab experiments have shown that facial attractiveness peaks, the waist-to-hip ratio shrinks, and body scent reaches its most appealing level (at least to heterosexual men). What's important to remember about these changes, though, is that they are subtle.

In a recent experiment in Europe, researchers photographed forty-eight young women, none of whom reported using birth control pills. (This is important, because the pill eliminates the effects of peak fertility on a woman's body by inducing a hormonal fake pregnancy.) The women were each photographed twice—once during a highly fertile phase of their cycle and once during a low-fertility phase. The photos were then shown to a group of men and women who were asked to choose the image they found more attractive. A little more than half the time—54 percent—those who were asked to rate the photos chose the photo taken when the woman was in the fertile phase of her cycle. This is somewhat better than chance (chance being 50 percent), but not much better.

> In humans, the signs of ovulation are all but invisible—yet somehow, we still know.

Indeed, fertility peaks during the menstrual period are so hard to detect that for years many sex researchers have concluded that in humans (as opposed to, say, baboons or other primates) signs of ovulation are all but invisible. From an evolutionary point of view, this is of profound importance: it suggests that male humans, unlike male baboons, can't tell the best time to mate.

But just because something is invisible doesn't mean we don't notice it. To demonstrate, researchers at the University of New Mexico recently tracked the menstrual cycles and the earnings of topless dancers who performed at "gentlemen's clubs" in the

Albuquerque area. In particular, the researchers were interested in how much money the women earned—not while they were on-stage performing for the entire audience, but while they were down on the floor engaging in intimate, one-on-one performances known as lap dances.

Lap dances are a staple in most strip clubs, accounting for the bulk of dancers' time and money. Typically, a dancer will perform on the elevated stage for only two or three songs every ninety minutes or so; the rest of the time she spends down on the floor soliciting customers for lap dances. In each dance, a man sits on a chair or couch, fully clothed, with his hands at his side (typically, he is not allowed to touch the dancer). The topless dancer sits astride his lap, either facing him (to display her breasts) or facing away from him (to display her backside). Usually, there is intense rhythmic contact between her pelvis and his. A lap dance usually costs $10 per song, or $20 per song if the dance is performed in the more private VIP area of the club. Typically, lap dances account for about 90 percent of a dancer's income. In all, the researchers tracked the women as they performed some fifty-three hundred lap dances over the course of two months.

The researchers found that the dancers' earnings were strongly related to their menstrual cycles. During fertile periods, normally cycling women (those not taking birth control pills) earned an average of $335 per five-hour shift. But during menstruation, their earnings plunged 45 percent, to $185 per shift.

Moreover, this pattern was remarkably consistent: *all* of the dancers made less money during their menstrual periods—whether they were on the pill or not. (Though those topless dancers on the pill

> **Topless dancers taking birth control pills consistently earned less than dancers who weren't on the pill—on average, about $80 less per shift.**

consistently earned less than those who weren't on the pill—on average, about $80 less per shift.)

The researchers' findings raised an interesting question: If clues about a woman's fertile phase are so hard to detect, as sex researchers have long believed, then how did the dancers' customers detect them? The men had no obvious hints; the women performed topless, but not bottomless. Nor did the dancers mention their fertility status to their customers. And in the dark, noisy confines of the topless bars, it was doubtful men could detect the fine changes in facial features noted by those in the European study mentioned above.

So how could they know? The New Mexico researchers were stumped. Somehow, the women were able to "signal" cues of their fertility state, and these cues influenced spending by their male customers.* It wouldn't be the first time researchers have demonstrated that the spending habits of men can be influenced by factors men can't see. Certain fragrances, for instance, have been shown to free up the wallet. In one experiment, the average amount of money spent by a man at a retail store that scented its air with a "male" fragrance was $55; but the average amount spent by men in a store that scented its air with a "feminine" fragrance was less than half as much—$23.

All this has important implications for understanding the sources of our mistakes. When we cast a vote or spend a dollar, we assume we do so for rational reasons. And if we later discover that our vote was miscast or our money misspent, we assume the explanation for that error must lie in the rational world. We do not think

---

*A possible explanation may lie in the women's voices. A recent study found that women who are at the peak time of fertility in their menstrual cycle may have changes in their voices that make them sound more attractive. This change may occur because the larynx changes its shape and size in response to hormones related to reproduction. For details, see Pipitone and Gallup (2008).

we voted for some guy because we made a judgment about him in under a second; we do not think we spent more in a store because it smelled good; and we don't think we gave the stripper a big tip because she may be about to conceive. But we did.

## It's Not the Wine; It's the Bottle

Similarly, we know that our minds form associations between certain traits and certain objects, even when we know that the object and the trait shouldn't necessarily be associated with each other. Take, for example, those two old sidekicks, price and quality. On one level, we all "know" that just because something costs more doesn't mean it's better than something that costs less. But deep down, we don't really believe that.

Take the case of expensive wine. Researchers from Stanford and the California Institute of Technology recently asked twenty volunteers to taste and evaluate five wine samples that were labeled according to price: $5, $10, $35, $45, and $90 a bottle. The volunteers were similar to many of us: they were moderate wine drinkers but not experts. And after tasting the wine, they replied as you or I might: they liked the expensive wine best.

But, as you might suspect, the researchers pulled a switcheroo on them. The $90 wine actually appeared twice—once in the $90 bottle and once in the $10 bottle. The same for the $45 wine: it appeared in the $45 bottle, but also in the $5 bottle. But the tasters never noticed; no matter what, they preferred the wine when it was in the more expensive bottle. And this was not simple snobbery at work. Brain scans showed that the higher-priced wines generated more activity in an area of the brain (the medial orbitofrontal cortex) that responds to certain pleasurable experiences. And when the drinkers drank the cheap

> People preferred wine when it was in the more expensive bottles. When they drank "cheap" wine their brains actually registered less pleasure.

wines? Their brains actually registered less pleasure from the experience.

The same process affects the way we rate medicines. In one shocking experiment (and I mean that literally), eighty-two men and women were asked to rate the pain caused by electric shocks applied to their wrists. In particular, they were asked to rate the pain at two different times: once after receiving the shock and again after they had taken a pill for their pain. Half of the people were informed that the pill they took was a newly approved prescription pain reliever that cost $2.50 per dose. The other half was informed that the pill cost just ten cents a dose. In both cases the medicines were dummy pills, known as placebos, with no pain-relieving qualities at all. The results: 85 percent of those using the expensive pills reported significant pain relief, compared with only 61 percent for those who took the cheaper pills.

## Color Counts

Price isn't the only thing that skews our judgment. Color can, too. For instance, previous studies have shown that a pill's color can also affect our perception of its potency. In one, people rated black and red capsules as "strongest" and white ones as "weakest."

To a degree, this association is understandable. There's a reason Johnny Cash wanted to be known as the Man in Black and not as, say, the Man in Bubble Gum Pink: we often equate blackness with power and strength. But these kinds of associations can backfire, causing us to make grave mistakes in judgment. For instance, years ago two researchers showed groups of trained referees one of two videotapes of the same aggressive play in a staged football game. In one tape, the aggressive team was wearing white uniforms; in the other tape, the aggressive team was wearing black. The referees who saw the black-uniformed version rated the play as being much more aggressive and more deserving of a penalty than those who saw the white-uniformed version. The referees "saw" what this common negative association—the color black—led them to expect to see.

"Okay," you may say. "That's fine in the lab. But what about the real world? What would happen to real teams that wear black? Would they be penalized more?"

Yes, they would. The same researchers compiled records for professional football and hockey teams during the 1970s and 1980s. For the National Football League, the records ranged from 1970 to 1986. (Nineteen seventy was the year the NFL merged with the now-defunct AFL to form one league with a single set of rules and a common group of referees.) For the National Hockey League, the records covered almost exactly the same period, ranging from 1970 to the 1985–86 season.

For both sports, the researchers found that teams that wear black uniforms have been penalized significantly more than average. This was especially true in hockey. Interestingly, during the sixteen-year study period, two teams—the Pittsburgh Penguins and the Vancouver Canucks—switched to black uniforms. The Penguins made the switch in the 1979–80 season, and the Canucks before the 1978–79 season.

> Teams that wear black uniforms have been penalized significantly more than average.

Can you guess what happened? Their penalty minutes increased. During the first forty-four games of the Penguins' 1979–80 season, when they wore blue uniforms, they averaged eight penalty minutes per game. But during the final thirty-five games of the year, when they wore black, their penalty minutes jumped to twelve per game.

## Why You Should Change Your Answer

We also know that we are powerfully influenced by first impressions—rightly or wrongly. For instance, have you ever taken a test, picked an answer, and then thought about changing it—but didn't?

If so, you're not alone. Most people tend to stick with their first answers. Three out of four college students, for instance, believe it is better to stick with their initial answer on a test rather than change it to one they think might be correct. Many college professors believe this as well. In one survey, only 16 percent of professors said they believed that changing an answer would improve a student's score; most believed that doing so would probably lower test scores. Even those in the test-preparation industry seem to believe this. Barron's *How to Prepare for the SAT*, for instance, admonishes students, under "Tactic No. 12," not to change answers capriciously. "In such cases, more often than not, students change right answers to wrong ones."

But, as is often the case in life, the majority is wrong. More than seventy years of research on answer changing shows that most answer changes are from wrong to right, and that most people who change their answers on a test improve their scores. This is true regardless of the type of test involved: multiple-choice or true-false, timed or not. One comprehensive review examined thirty-three studies of answer changing. In not one were test takers hurt, on average, by changing their answers.

> Most people who change their answers on a test improve their test scores.

And yet the myth of sticking with first answers persists to this day. Studies have shown that even after students are told of the research on answer changing, they *still* tend to stick with their first answers.

"The fact that that isn't true is very surprising to a lot of people," says Justin Kruger, a professor at New York University's Stern School of Business who has extensively studied people's first instincts. "It's really counterintuitive to educators and students and test takers themselves. People generally have this lay belief that as a general rule you should stick with your first instinct. And the fact of the matter is there isn't a lot of evidence to support that."

## The Role of Regret

What there is an increasing amount of evidence to support is the subtle but powerful role in our decision making that is played by emotions, especially the emotion of regret. We have all faced situations that produced outcomes we regret: bad marriages, bum cars, real estate we couldn't unload. But some of the choices we make produce more regret than others; and the difference in levels of regret helps explain why we often cling so tightly to our first instincts.

As a general principle, people feel more responsible for their *actions* than they do for their *inactions*. If we are going to err at something, we would rather err by failing to act. That's because we tend to view inaction as a passive event—we didn't *do* anything. And since we didn't do anything, we feel less responsible for the outcome that follows. This was illustrated in a series of experiments conducted by Kruger and his colleagues. They looked at the test-taking practices of more than sixteen hundred college students. Not surprisingly, the researchers found what others before them have found: test takers who changed their answers usually improved their score. In fact, when all the changed answers were counted and analyzed, changes from wrong to right outnumbered changes from right to wrong by a margin of two to one.

> If we are going to err at something, we would rather err by *failing* to do something.

But more important is what the students revealed in follow-up interviews: the prospect of changing a right answer to a wrong one filled them with much more regret than the prospect of *failing* to change a wrong answer to a right one. In short, doing nothing was less regrettable than doing something—even though, in both cases, they'd end up with the wrong answer.

### The "Monty Hall" Problem

On one level, this difference is not a surprise. This phenomenon is so commonly encountered by researchers that they have dubbed it the "Monty Hall" problem, named after the venerable host of the long-running television game show *Let's Make a Deal.* If you're old enough to remember the program (it went into reruns in the mid-1970s), the show's contestants were invariably confronted with the same dilemma: Do you take what's behind Door No. 1, or do you keep what you have?

But on another level, the student's answers on regret raised a novel question: Does a difference in regret lead to a difference in the way we remember certain events? For instance, pretend you're taking a test. You have two options: in one, you go against your first instincts, change your initial answer, and get the problem wrong; in the other, you stick with your original choice, but still get the answer wrong. Will you remember both events equally well?

### Some Mistakes Are More Memorable than Others

To test this idea, Kruger gave students questions from portions of two actual college entrance exams: the Scholastic Aptitude Test, which is used at the undergraduate level, and the Graduate Record Examination, which is used for admission to graduate programs. Once students had narrowed down their answers on a given question to two possibilities, Kruger asked them to note which of the two had been their first guess. Then, about a month later, he asked them detailed questions about how they answered the items on the tests.

He found that students showed a marked memory bias—and that this bias was caused by regret. In analyzing the answers on the students' tests, Kruger found that the students were more likely to get the problem wrong if they stuck with their first answer. But this, interestingly, is not what the students remembered. When Kruger asked them how often they had switched their answer and got it wrong, they overestimated. When he asked them how often they stuck with their first answer and got the problem wrong, they underestimated.

Bottom line: the students remembered sticking with the first answer as being a better strategy than it actually was. It is this memory bias, says Kruger, that helps explain why people continue to believe that sticking with the first answer is a more effective strategy than it is.

"The paradox is even though the actual outcome of your answer changing suggests that you should be doing more of it," he said, "your memories of it suggest the very opposite."

Chapter 4

# We Wear Rose-Colored Glasses

What's the world's clumsiest and goofiest thing to do?

No doubt, the list of candidates is endless. But if you're the billionaire casino mogul Steve Wynn, your nominee would have occurred on or about September 30, 2006. That's when some famous friends of his, including the broadcaster Barbara Walters and the writer Nora Ephron, were visiting Wynn in his Las Vegas office. Wynn, a noted art collector, took the opportunity to show off one of his most treasured possessions—a 1932 portrait of Picasso's mistress, Marie-Thérèse Walter, titled Le Rêve (The Dream).

The erotic painting is notable for a number of reasons, not the least of which is that the head of Marie-Thérèse is divided into two sections, one of which is a penis. (If you've been to Las Vegas, maybe you've seen it—the painting used to hang in the museum at the Bellagio Hotel and Casino when Wynn owned the Bellagio.)

This aside, the portrait was also among the most valuable works of art in the world. Just a day before receiving his visitors, Wynn had agreed to sell the painting to another billionaire, the hedge fund executive Steven Cohen, for the whopping sum of $139 million. This would have been $4 million more than the previous record price paid for a work of art.

But just as Wynn was showing off the portrait to his friends, he gestured with his right hand—and put his elbow through the painting.

"Oh shit, look what I've done," said Wynn, according to Ephron, who provided an account of the incident on a blog.*

But that's not what Wynn remembered saying. In an interview published a few months later (after Wynn had filed an insurance claim with Lloyd's of London), he remembered his own language as being somewhat more delicate.

"I just turned and said, 'Oh my God! How could I have done this?' "

The difference in the wordings is small but telling. In remembering our own actions, we all tend to wear rose-colored glasses. Without intentionally trying to distort the record, we are prone to recalling our own words and deeds in a light more favorable than a watching blogger might record.

## We Remember Our As

To demonstrate, let me ask you a question that you can answer objectively—but only if you kept all of your old report cards. Here it is: How'd you do in high school?

The answer is: probably not as well as you remember—at least not if students at Ohio Wesleyan are any guide. In one study, students at OW were asked to recall their high school grades. Then the researchers checked the students' responses against the actual transcripts. No less than 29 percent of the recalled grades were wrong. (This, at least, was better than German students. Given a similar

*Ephron's account, which is predictably hilarious, can be found at the Huffington Post at www.huffingtonpost.com/nora-ephron/my-weekend-in-vegas_b_31800.html. Wynn would later claim the damage had reduced the value of the painting by $54 million; he called his own actions "the world's clumsiest and goofiest thing to do." See Zambito (2007).

test, they did even worse: 43 percent wrong.) Keep in mind, this is not ancient history the students are being asked to recall; these are college freshmen and sophomores being asked about their grades in high school just a few years earlier.

What's more, the students' errors weren't neutral. Far more grades were shifted up (recalling an A instead of a B) than down. Not surprisingly, perhaps, students had a better memory for good grades than for bad. The recall accuracy for As was 89 percent; for Ds it was 29 percent. (Researchers threw out the Fs.) Nor were the errors isolated. Overall, seventy-nine of the ninety-nine students inflated their grades. (Too few students deflated their grades to allow meaningful generalizations.)*

> The recall accuracy for As was 89 percent; for Ds it was 29 percent.

The experience at Ohio Wesleyan is hardly unique. Time and again, people have been shown to reconstruct their memories in positive, self-flattering ways. Parents, for instance, have been shown to remember their parenting methods as being far closer to what expert opinion would prescribe than they actually were. Likewise, gamblers remember their wins more kindly than they do their losses. And nearly all of us, as we'll see, misremember a key aspect of our sexual past.

This inclination is so powerful that it extends even to the way we *see* ourselves. In a recent series of experiments, Nicholas Epley of the University of Chicago and Erin Whitchurch of the University of Virginia demonstrated that we recognize our own faces as being more physically attractive than they actually are. In the experiments, people were asked to identify pictures of them-

---

*A more recent study, performed on students at Northeastern University in Boston, found that self-reported grade point averages "were significantly higher than actual GPAs recorded by the registrar." Only 10 percent of the students remembered their grades as being lower. See Gramzow, Willard, and Mendes (2008).

selves from a lineup of distracter faces. Participants identified their own portraits much more quickly when their faces were computer enhanced to be 20 percent more attractive. Not only that, they were also likelier, when presented with images of their faces that were prettier, homelier, or left untouched, to call the prettier version of their face the genuine, unretouched face. (They showed no such tendency, by the way, when viewing the faces of strangers.)

Indeed, the tendency to see and remember in self-serving ways is so ingrained—and so subtle—that, like many of the other errors discussed in this book, we often have no idea we're doing it. The Princeton Nobel laureate Daniel Kahneman, in an interview, observed this some time ago.

"I mean, the thing that is absolutely the most striking is how seldom people change their minds," he said. "First, we're not aware of changing our minds even when we do change our minds. And most people, after they change their minds, reconstruct their past opinion—they believe they always thought that."

Believing that we always thought this or that might be harmless enough if such beliefs were confined to our past *opinions*. If we remember ourselves as being better parents than we actually were, for instance, so what? Maybe there will be some rolled eyes over the egg salad at the family reunion, but nothing more. But what about past *facts?* When people are really put under the spotlight—not to mention under oath—do their memories of their own actions still tend to be self-serving?

### Your World Revolves Around You

Some of you may recognize the name of John Dean. He's the author of the book *Conservatives Without Conscience* and appears from time to time on cable TV news shows as a political commentator. Dean came to fame in the 1970s as counsel to then president Richard M. Nixon during the Watergate scandal. In June 1973, Dean testified to the Senate Watergate investigating Committee about conversations he had with Nixon in the Oval Office. As it later, famously, turned out, those conversations were tape-recorded. This proved to be a boon for Ulric Neisser, a psychologist from Cornell University and one of the world's leading authorities on human memory. The recordings gave Neisser an opportunity that almost never occurs in the real world: the chance to compare a memory of a conversation with the conversation itself.

"What could be better?" asked Neisser, who is now retired.

Neisser found that Dean's memory, like those of the Ohio Wesleyan students, contained systematic distortions and that those distortions tended to be in Dean's favor. Many of the distortions, for instance, reflected Dean's own self-image, with Dean tending to recall his role as being more central than it was. But, more important, Dean's memory was frequently wrong—not only in the recollection of details of the conversation, but even regarding the gist of the conversation. In fact, Neisser concluded, "comparison with the transcript shows that hardly a word of Dean's account is true."

Below is a portion of Dean's prepared statement to the committee, part of which described a meeting with Nixon on September 15, 1972. The statement is quite voluminous—245 pages in all—about the length of this book. You and I would be unlikely to dictate a manuscript that long from memory, and neither did Dean. To prepare a statement of this length, Dean engaged in an act of reconstruction: he consulted a newspaper clipping file that he kept from the date of the Watergate break-in until the time of the Watergate hearings. He used the file as a kind of memory aid, going through each article outlining what had happened, then placing himself back in the situation described in the article. It was through this process, he told the senators, that he was able to fashion such a detailed recollection.

Again, it's worth remembering that Dean, like the students at Ohio Wesleyan, was recalling not ancient history but events that occurred just nine months before his testimony. It's also worth noting that Dean was renowned for his ability to recall details. Some writers even dubbed him "the human tape recorder," and Dean seemed to encourage such an impression. During the hearings, for instance, Dean was specifically asked by one senator about how, exactly, he was able to recall so much information in such detail.

"Well, Senator," Dean said, "I think I have a good memory. I think that anyone who recalls my student years knew that I was very fast at recalling information, retaining information. I was the type of student who didn't have to work very hard in school because I do have a memory that I think is good."

Here, then, is the excerpt from Dean's statement:

On September 15 the Justice Department announced the handing down of the seven indictments by the Federal Grand Jury investigating the Watergate. Late that afternoon I received a call requesting me to come to the President's Oval Office. When I arrived at the Oval Office I found Haldeman and the President. The President asked me to sit down. Both men appeared to be in very good spirits and my reception was very warm and cordial. The President then told me that Bob—referring to Haldeman—had kept him posted on my handling of the Watergate case. The President told me that I had done a good job and he appreciated how difficult a task it had been and the President was pleased that the case had stopped with Liddy. I responded that I could not take credit because others had done much more difficult things than I had done. As the President discussed the present status of the situation I told him that all I had been able to do was to contain the case and assist in keeping it out of the White House. I also told him there was a long way to go before this matter would end and that I certainly could make no assurances that the day would not come when this matter would start to unravel.

A review of the transcript, Neisser found, shows that Nixon did not say any of the things attributed to him here. He didn't ask Dean to sit down. He didn't say Halderman had kept him posted. He didn't say Dean had done a good job (at least not in that part of the conversation). And he didn't say anything about Liddy or the indictments. Nor had Dean himself said the things he later describes himself as saying, as can be seen from the actual excerpts from the tape-recorded conversation below:

President: Hi, how are you? You had quite a day today, didn't you? You got Watergate on the way, didn't you?

Dean: We tried.

Haldeman: How did it all end up?

Dean: Ah, I think we can say well, at this point. The press is playing it just as we expected.

Haldeman: Whitewash?

Dean: No, not yet—the story right now—

President: It is a big story.

Haldeman: Five indicted plus the WH former guy and all that.

Dean: Plus two White House fellows.

Haldeman: That is good; that takes the edge off the whitewash, really. That was the thing Mitchell kept saying, that to people in the country Liddy and Hunt were big men. Maybe that is good.

President: How did MacGregor handle himself?

Dean: I think very well. He had a good statement, which said that the Grand Jury had met and that it was now time to realize that some apologies may be due.

Haldeman: Fat chance.

Dean: Get the damn (inaudible).

Haldeman: We can't do that.

President: Just remember, all the trouble we're taking, we'll have a chance to get back one day. How are you doing on your other investigation?

The conversation is briefly interrupted when Nixon takes a phone call. As soon as he hangs up, Dean reminds him of how well things are going:

Dean: Three months ago I would have had trouble predicting there would be a day when this would be forgotten, but I think I can say that 54 days from now [that is, on Election Day in November], nothing is going to come crashing down to our surprise.

President: That what?

Dean: Nothing is going to come crashing down to our surprise.

I contacted Dean to ask him what he made of Neisser's assessment. In an e-mail, he told me that he was not familiar with Neisser's analysis of his Senate testimony. Nonetheless, Dean added, it would not be the first to compare his testimony with the tapes. "Typically, I have found that those making this drill ignore my statements made during my testimony where I explained that all I was—or anyone is—capable of doing was characterizing earlier conversations, and making clear neither my memory—nor anyone else's—is a tape recorder. Since I believed I had been taped by Nixon, I actually under-testified about what I thought I did recall."

So I sent him a copy of Neisser's article, hoping for a more detailed analysis. In return, I got a very short note. "I believe Neisser has distorted both my testimony and the tapes," he wrote. "The problem is, it would take a lot of time that I don't have to marshal all the information." After that, Dean stopped responding to my inquiries.

### Hindsight Isn't Twenty-Twenty

Nevertheless, Dean's testimony underscores an important point: hindsight isn't twenty-twenty; it isn't even close. In fact, one of the most significant sources of human error is hindsight bias. Basically, hindsight bias comes down to this: knowing how things turned out profoundly influences the way we perceive and remember past events. This is true no matter how trivial they may be. It could be the 1975 Super Bowl or Grandma's colostomy or the decision to neuter Rover; knowing how the event turned out alters our recollection of it.

> Knowing how things turned out influences the way we perceive and remember past events, giving the outcome an appearance of inevitability.

Even historians are prone to this error. It is much easier after an event—whether it is the Battle of Gettysburg or the bombing of Pearl Harbor—to sort the relevant factors from the irrelevant. Those who write of these events will almost invariably give the outcome an appearance of inevitability. But this type of compelling narrative is achieved by suppressing some facts at the expense of others—a process known as creeping determinism.

Near the conclusion of her influential history of the attack on Pearl Harbor, for instance, the noted military analyst Roberta Wohlstetter had this to say: "After the event, of course, a signal is always crystal clear; we can now see what disaster it was signaling, since the disaster has occurred. But before the event it is obscure and pregnant with conflicting meanings."

Experiments on hindsight bias have shown not only that people exaggerate what they would have known at the time but that they actually misremember what they knew. This is especially true if it turns out that they were wrong in the first place.

Much of the pioneering work on hindsight bias was done in the 1970s by Baruch Fischhoff, then at Hebrew University in Jerusalem and now at Carnegie Mellon University in Pittsburgh. Fischhoff, like Neisser, got a great chance to study hindsight bias from Richard Nixon. In 1972, Nixon took not one but two historic trips abroad: one to China, the other to the Soviet Union. The press was filled with speculation about what Nixon would or would not accomplish on these trips. But for Fischhoff, these issues were secondary. What counted is that the trips would provide a perfect real-world test of hindsight bias.

Before Nixon left on his trips, Fischhoff and his colleague Ruth Beyth asked a group of Israeli students to estimate how likely certain outcomes were. For instance, they might be asked, "How likely is it that Nixon will meet with Chairman Mao?" Or, "How likely is it that Nixon will visit Lenin's tomb?" In short, they asked the students to do what bookies do every day: lay odds. If the students thought there was no chance the event would happen, they were to assign it a probability of 0 percent. If they thought the event was sure to happen, they were to assign it a probability of 100 percent.

After Nixon completed his trips to Beijing and Moscow, Fischhoff would periodically ask the same students to remember, as best they could, their original predictions. He also asked them a key second question: Did they believe that the event about which they had been asked to make a prediction had, in fact, occurred? For instance, if they had predicted that Nixon would meet with Mao, did they believe that Nixon had, in fact, met with Mao?

Fischhoff found that over time, the people in his study mis-remembered their original predictions, just as the students at Ohio Wesleyan had misremembered their high school grades. And, just like the memories of the Ohio students (not to mention John Dean), the memories of the Israeli students tended to make them look smarter than they were. In particular, Fischhoff found that if a person believed that the predicted event had, in fact, happened, then he exaggerated the odds made in the original prediction. For instance, if the original prediction had been, say, 30 percent, then it might be remembered as having been 50 percent. But if the students believed the event did *not* happen, they erred in the opposite direction: they low-balled their original prediction, so that an initial guess of 50 per-cent became, say, 30 percent. In either case, the effect was the same: those in the study remembered their own actions in a way that made them appear more foresighted than they actually were. In fact, Fischhoff titled one of his papers "Hindsight ≠ Foresight."

### Remembering Our Sex Partners

Men routinely report more lifetime sex partners than women do.

Our everyday lives are littered with a simi-lar kind of misremembering. Take sex, for instance. You probably remember, over the course of your life, how many people you have had sex with. And you no doubt think the number you remember is accurate. But what if you asked your friends to recall the same information about *their* sex lives? Chances are the answer from your male friends would be much different from that of your female friends. Professional pollsters have cer-tainly seen this. On national surveys, men routinely report up to four times as many lifetime partners of the opposite sex as women do—even though they should each report the same number since each new sex partner for a man is also, obviously, a new sex partner for a woman.*

Why there should be such a big gap is unknown. But what seems to be clear is this: men and women don't intentionally misrepresent their sexual histories. Some people, of course, do lie to pollsters about their sex lives. But when men and women are asked about other, equally intimate aspects of their sex lives, they've been found to give answers that are similar to each other. For instance, men and women are equally likely to acknowledge having engaged in oral or anal sex. They also give similar answers when asked about how often they have sex or how long their sexual encounters last. So it seems unlikely that they would tell the truth about these parts of their sex lives but lie when it comes to accounting for the number of sex partners they've had.

Instead, the most likely explanation seems to be that the men and women simply misremembered. Moreover, like the Ohio Wesleyan students, they misremembered in a particular direction: the one that appears most self-serving (and stereotypical). Men exaggerate their number of sex partners; women downplay theirs.

### Why Gamblers Are So Confident

A similar kind of revision affects the memories of people who like to gamble, especially on sporting events like football games. If you know people like this, ask them to tell you about some of their wins—then ask them to tell you about some of their losses. Listen

*In case you're keeping tabs, the most recent U.S. data show that the median number of lifetime female sexual partners for men is seven; the median number of lifetime male sexual partners for women is four. But, as with so much in life, the number of lifetime sex partners depends on how rich we are. A separate survey shows that the average number of lifetime sex partners in wealthy countries is ten; in poor countries, it's six. See Fontes and Roach (2007).

closely to what they say. Odds are they won't recall them in the same way. Instead, gamblers typically tend to accept their wins at face value but explain away their losses.

Years ago the Cornell professor Tom Gilovich, who was then at Stanford, spent his dissertation tracking the betting practices of experienced gamblers in professional football and college basketball games.

"I looked around and had a bunch of these people in my life," Gilovich told me. "And I always wondered, 'Why do these people have so much confidence?' They get beaten down and come right back, which is admirable in some ways. But in other ways, lives can be ruined by this."

So he set out not only to track whether gamblers won or lost but to learn what they *thought* about those wins and losses. To that end he handed the gamblers in his experiments tape recorders and asked each of them to record their thoughts on their winning and losing bets. The tapes were transcribed and then analyzed. Gilovich found that when gamblers were right, they tended to offer bolstering comments about just how right they were—"I knew it would happen," or words to that effect. But when they were wrong, they tended to minimize their error by offering "undoing" comments about how the game should have turned out differently. In these cases the gamblers would often blame the outcome on a fluke event, like a fumble in the fourth quarter. To them, a loss wasn't really a loss; it was a near win. In either case, the effect of the bolstering and undoing comments was largely the same: foresight became better in hindsight.

> In a gambler's hindsight, a loss isn't really a loss—it's a near win.

### We're Shallower than We Realize
The type of misremembering involved in hindsight carries with it a significant, though unstated, implication: we don't always know

when we're being biased. In this sense, we don't mean "bias" in the way that people usually define the word—as some form of overt prejudice against certain types of people or ideas. Instead, we mean something much more nuanced—a small nudge to our judgment that occurs without our being aware of it.

Think back to the section on first impressions. Remember what happened when researchers showed pictures of politicians to the subjects in their experiments? Without studying the politicians' position papers or listening to a single word of their speeches, the subjects were able to draw amazingly quick inferences about the politicians' competence; all they had to do was glance at the politicians' faces.

Moreover, this snap judgment was no fleeting impression; it appeared to influence how the participants in the study would actually vote. When the researchers conducted simulated voting studies, they found that additional information about the politicians that voters would normally gather during the course of a campaign diluted—but did not eliminate—the impact of the initial impression. The actual voting preferences of the participants were anchored on their split-second inferences of competence from facial appearance.

This finding has profound implications for the rationality not only of voting preferences but of other preferences, too; many of the consequential decisions we make may be shallower than we would like to believe. Indeed, we may not even know that we have made a decision. In the case of the voters above, the researchers concluded that because such judgments were made so quickly, "their influence on voting decisions may not be easily recognized by voters." In other words, voters' opinions may have been biased; they just may not *know* they have been biased.

This may be hard for many of us to swallow because people tend to pride themselves on their impartiality.

"People always think they're not biased—even when you can

> No matter the issue, "people always think they're not biased— even when you can document statistically that there's a huge bias."

document statistically that there's a huge bias," says George Loewenstein, a professor at Carnegie Mellon and one of the nation's leading experts on the role that bias can play in shaping our judgments. And if we don't know we've been biased, eliminating the mistakes that stem from that bias can become much more difficult.

### Doctors Are Under the Influence

Consider the case of a doctor who writes you (or anyone else, for that matter) a prescription. Prescription drug use in the United States has been soaring. Between 1995 and 2000, annual retail spending on prescription drugs in the United States more than doubled, to $132 billion. Today, nearly half of all Americans take at least one prescription drug.

But much of this prescribing appears to be unnecessary. One study reviewed nearly thirty articles that looked at doctors' prescribing practices. It found, among other things, that interactions between doctors and pharmaceutical companies led to what the authors of the study politely termed "non-rational prescribing." Between 1989 and 2000, federal regulators approved more than a thousand new drugs. If all these new drugs represented significant advances, that would be one thing. But they didn't. The U.S. Food and Drug Administration determined that 76 percent of all new drugs approved during this time were no more than moderate innovations over existing treatments; many were simply a modification to an older product with the same ingredient.

But that didn't stop drug companies from charging a lot for them. In 2000, on average, these standard-rated new drugs cost nearly twice as much as existing drugs prescribed for the same condition. The additional revenue generated by these drugs has helped pharmaceutical companies spend more than $8,000 per doctor per year in the United States to promote the use of their drugs.*

Does this largesse influence the judgments of physicians? Doctors typically say no; surveys consistently show that doctors say they are not biased by financial arrangements with drug companies. But much research suggests otherwise. One survey, for instance, found that 84 percent of doctors thought their colleagues were influenced by gifts from pharmaceutical companies. But as for themselves? Only 16 percent thought they were similarly influenced.

> One survey found that 84 percent of doctors thought their colleagues were influenced by gifts from drug companies. But as for themselves? Only 16 percent thought they were similarly influenced.

### Disclosing a Bias Doesn't Cancel the Bias

The bias of physicians who take money from drug companies poses a classic conflict of interest: Is the doctor doing what's in the doctor's best interest, or is he doing what's in the patient's best interest? Typically, when these kinds of conflicts arise, the proposed cure is to disclose the conflict. Stock analysts, for instance, are usually required to disclose whether they have a stake in the companies they cover. And lawyers are required to tell clients whether they represent other clients whose interests may be adverse to their own. Some doctor groups, too, now insist that their members tell patients about financial conflicts. A new set of professional standards from the American Academy of Orthopaedic Surgeons, for instance, mandates that surgeons in the group begin disclosing to patients

---

*In 2008, the pharmaceutical industry said it would adopt voluntary guidelines to ban certain gifts to U.S. doctors. See Harris (2008).

any financial ties they have with industries that relate to the patients' treatment.*

In each of these cases, the same principle is at work: let the customers decide whether there's a bias, and if they think there is, let them decide how big it is and how much to correct for it.

But do such disclosures work? That's the question that George Loewenstein wanted to tackle. To find out, he and colleagues at Carnegie Mellon created a game not a whole lot different from the one most of us play when we invest in the stock market. Before investing our money, we usually need to decide what a given stock is worth: Ten dollars a share? Twenty dollars a share? To make that decision, we often rely on the advice of someone who, presumably, has more knowledge about the stock market than we do: someone like, say, a financial adviser.

In Loewenstein's game, his investors faced the same kind of decision. But instead of trying to value a share of stock, they were asked to guess the value of a jar full of coins. To help them, Loewenstein provided paid advisers who, indeed, had additional information about the value of the coins in the jar. It was the advisers' job to help the investors guess how much the coins in the jar were worth.

For our purposes, the important part of the experiment is in how these advisers were paid. In some cases, the adviser's pay was based on the accuracy of the client's guess: the more accurate the guess, the bigger the adviser's payday. But in other cases, the adviser's pay was based on how *high* the client's guess was. The higher the client's guess, the more money the adviser made. This situation

*Patients can already find out if their surgeon is paid by any of the five biggest orthopedic-equipment makers by visiting the companies' Web sites. The companies were recently required to make this information public as part of an agreement with federal prosecutors in New Jersey, which investigated kickbacks in the industry. See Armstrong (2007).

When financial
advisers' conflict of
interest was disclosed
to investors, the
advisers gave worse
advice than when it
wasn't disclosed.

represented a clear conflict between the interest of the adviser and the interest of the client: the client expected to receive an *accurate* estimate from the adviser; but it was in the adviser's financial interest to provide a *high* one instead. In some of these instances, this conflict of interest was disclosed to the clients; in others, it was not.

Then the game began. If you invest in the stock market, you will probably find the results interesting, if not encouraging. First, the conflict of interest had the effect you might expect: it raised the estimates given by the advisers. But disclosing the conflict made this effect even worse. Loewenstein found that the advisers consistently inflated their estimates of the value of the coins in the jar, usually by significant amounts. For example, when the advisers were paid based on the accuracy of their clients' guesses, their estimates were relatively low—just over $16. But when the advisers' compensation scheme shifted and they were paid based on how high the client's guess was, they bumped up their estimates to just over $20, on average. When the advisers were paid based on how high their clients guessed *and* this conflict was disclosed to the client, their estimates were highest of all: just over $24.

The second—and in some ways more important point—concerns the reaction of the investors. When the conflict of interest was disclosed to them, the investors did, indeed, discount the advice they received from their advisers—but not by nearly enough. On average, the investors discounted their advisers' estimate by $4—but the advisers had increased it by an average of $8! In other words, the investors had discounted the advice by only half as much as they should have.

When the conflict of
interest was disclosed
to them, investors
did, indeed, discount
the advice they
received from their
advisers—but not by
nearly enough.

What could account for these results? A couple of things. The easiest and most obvious answer is that the advisers, figuring their advice would be discounted once their conflict was known, tacked on a couple of extra dollars to their estimates to make up for the discounting they expected their clients would apply to their advice.

> When people demonstrate that they are not corrupt in some way, they are actually more likely to display exactly this corruption on subsequent tasks.

A less obvious but more interesting explanation has to do with an effect that psychologists call moral- or self-licensing. In laboratory experiments, researchers have shown that when people demonstrate that they are not corrupt in some way, they are actually *more* likely to display exactly this corruption on subsequent tasks. Researchers at Princeton, for instance, recently showed that once people have established credentials as being non-prejudiced, they actually reveal a *greater* willingness to express politically incorrect opinions.

### The "Hey, I Warned You" Principle

We saw an example of this during the 2008 presidential campaign. Longtime civil rights leader Jesse Jackson, while discussing Barack Obama during a taping of a Fox News program, was overheard using the N-word in off-air remarks. (Jackson, famously, also used an ethnic slur years earlier, describing New York City as "hymie-town.")

An even more telling example is provided by the nation's tobacco companies. After the federal government, in 1965, required tobacco companies to put warning labels on cigarette packages, consumer advocates hailed the move as a great victory. But since then, the tobacco industry has fended off smokers' lawsuits by citing the warning label as evidence that smokers should have known the risks involved in smoking. In effect, the tobacco companies said, "Hey, I warned you."

Disclosures of conflicts of interest function in much the same way. They tell people, "Hey, I warned you." And just as cigarette companies apparently felt that the warnings gave them a moral license to continue selling products that killed people, Loewenstein's advisers apparently felt they had a similar license to pursue their own self-interest, even at the expense of their clients. ("Hey, I warned you.")

Although Loewenstein's findings were confined to the lab, he thinks the same forces are "pretty pervasive" in the real world.

"Suppose you are a casual investor and you find out that IBM is getting consulting services from its auditor," says Loewenstein. "Or suppose your doctor recommends an X-ray, and then says he has an interest in the X-ray facility. In neither of those cases would you have a clue what to do. Would you not go get the X-ray? Is IBM worth half as much? Ten percent less? Five percent less?"

Usually, he says, people don't know what to do with the information.

"So what they do is ignore it."

That's why disclosing a bias doesn't cancel its effects. The best way to do that, says Loewenstein, is to eliminate the bias in the first place. In any event, the bottom line of Loewenstein's experiments was unchanged: when the conflict of interest was disclosed, the advisers made more money—and their customers made less.

> Usually, people don't know what to do with information about conflicts of interest. "So what they do is ignore it."

*Chapter 5*

# We Can Walk and Chew Gum—
# but Not Much Else

The next time you fly, remember Captain Robert Loft. Captain Loft, the pilot of Eastern Airlines Flight 401, was making his final approach to Miami International Airport when he noticed something was wrong. He had put the landing gear down, but the indicator light didn't come on. So he circled around, leveled off at two thousand feet, and decided to have a look.

He couldn't figure it out, so he called in the first officer. The first officer had no luck, so they called in the flight engineer. As chance would have it, a mechanic from Boeing happened to be flying that day and was seated in the jump seat of the cockpit. So they asked him, too, to have a look. Pretty soon, nobody was flying the plane. It went lower and lower. Suddenly, the captain was seized with a realization.

"Hey!" he shouted. "What's happening here?"

Those were his last words. Five seconds later, the plane plowed into the Everglades and burst into flames. Ninety-nine people, including Captain Loft, were killed. A study of the crash later determined that the crew had become so engrossed in the task that they had lost awareness of their situation—all because of a $12 light-bulb.

The crash wasn't a fluke. The experience of flying a perfectly good airplane into the ground is so common that an engineer from Honeywell coined a term for it: "Controlled Flight into Terrain," or CFIT for short. Despite a number of technological innovations, CFIT remains one of the most lethal hazards in aviation. Forty percent of aircraft accidents and well over half of all aircraft fatalities have been attributed to CFIT. Since 1990, no other type of airline accident has taken more lives.

> The experience of flying a perfectly good airplane into the ground is so common that engineers coined a term for it: "Controlled Flight into Terrain," or CFIT for short.

An obvious question, of course, is: What would cause pilots to do such a thing? A few years ago, the U.S. Air Force examined this very question. CFIT accidents were exacting a terrible toll on the Air Force: between 1987 and 1998, CFIT accidents had accounted for 190 fatalities, 98 lost aircraft, and total costs of $1.7 billion. When the Air Force scrutinized the factors behind such accidents, it found that most of them had something in common: in more than half of the cases, the flight crews had, like Captain Loft, lost awareness of the situation inside the cockpit. They had become so engrossed in what they were doing that they had lost the ability to fly the airplane; one reason for this was what the Air Force called "task saturation"—that is, trying to do too many things at one time.

## We're Not Really Multitasking

Down here on the ground, we don't experience CFIT. But we do experience something close: multitasking. Whether we're on foot or behind the wheel of a car, our attention, like that of Captain Loft, is continually being divided by the tasks we try to juggle: listening to our iPods, talking on our cell phones, tapping away on our BlackBerrys.

"Multitasking" is a term cribbed from the computer world; it describes a technique by which a computer can split up its work into many processes or tasks. This allows us to, say, run Microsoft Word while downloading something from the Internet. Most of us think our brains can work in the same way. Indeed, multitasking has become the hallmark of the modern workplace. Gloria Mark, a professor at the University of California, Irvine, who studies multitasking in the workplace, recently conducted a field study of employees at an investment management company on the West Coast. She and a colleague watched as the workers went about daily tasks in their cubicles; they noted every time the workers switched from one activity to another—say, from reading an e-mail that popped up in their inbox to making a phone call to jotting something down on a Post-it note. They found that the workers were frequently interrupted—on average, about twenty times an hour. This means the employees were, on average, able to focus on one task for no more than about three minutes.

But multitasking is one of the great myths of the modern age. Although we think we are focusing on several activities at once, our attention is actually jumping back and forth between the tasks. Not even a computer, by the way, can multitask; it actually switches back and forth between tasks several thousand times per second, thus giving us the illusion that everything is happening simultaneously.*

> Not even a computer multitasks; it switches back and forth between tasks several thousand times per second, thus giving the illusion that everything is happening simultaneously.

*Some modern computers do have multiple processors, and these truly do allow a computer to perform multiple tasks at the same time; like a person with two or more heads, each processor can work (or perform) independently. But in the old days, when the term "multitasking" was coined, computers had just a single processor.

Our minds provide us with the same illusion, but not, unfortunately, the same results. There is no such thing as dividing attention between two conscious activities. Under certain conditions we can be consciously aware of two things at the same time, but we never make two conscious decisions at the same time—no matter how simple they are. Sure, you can walk and chew gum at the same time. And you can drive and talk to a passenger at the same time, too—but only after so much practice that the underlying activity (walking or driving) becomes almost automatic. But we don't practice most of our day-to-day activities nearly enough for them to become automatic. The next time you're at a restaurant, for instance, try carrying on a conversation with your dinner guests while trying to figure the tip on the bill.

## Multitasking = Forgetting

Indeed, the gains we think we make by multitasking are often illusory. That's because the brain slows down when it has to juggle tasks. We gain nothing, for instance, by ascending the stairs two steps at a time if the additional effort slows us down so much that we end up taking as long to climb them as we would if we had taken them just one step at a time. In essence, this is what often happens when we try to perform two mental tasks simultaneously. In one experiment, researchers asked students to identify two images: colored crosses and geometric shapes, like triangles. Seems simple enough, right? When the students saw colored crosses and shapes at the same time, they needed almost a full second of reaction time to press a button—and even then they often made mistakes. But if the students were asked to identify the images one at a time—that is, the crosses first, *then* the forms—the process went almost twice as quickly.

Switching from task to task also creates other problems. One of them is that we forget what we were doing—or planned to do. That to-do list in our brains is known as working memory; and it keeps track of all the short-term stuff we need to remember, like the e-mail address someone just mentioned to us. But the contents of our working memory can evaporate like water in a desert; after only about two seconds, things begin to disappear. And within fifteen seconds of considering a new problem, researchers have shown, we will have forgotten the old problem. In some cases, the forgetting rate can be as high as 40 percent. This obviously presents the potential for big mistakes—especially if you're, say, an air-traffic controller.*

> Switching from task to task causes us to forget what we were working on in the first place; in some cases, the forgetting rate can be as high as 40 percent.

Another cost is downtime. When we're working on one thing and are interrupted to do another thing, it takes us a while to refocus on what we were originally working on. Workplace studies have found that it takes up to fifteen minutes for us to regain a deep state

> Workplace studies have found it takes up to fifteen minutes for us to regain a deep state of concentration after a distraction such as a phone call.

of concentration after a distraction such as a phone call. These findings square with what researchers found when they looked at the work habits of employees at Microsoft. In that study, a group of Microsoft workers took, on average, fifteen minutes to return to serious mental tasks, like writing reports or computer codes, after responding to incoming e-mails. Why so long? They typically strayed off to reply to other messages or browse news, sports, or entertainment on Web sites.

So long as such distractions are confined to our cubicles, most of us are probably safe. But in the real world, researchers are dis-

---

*When the U.S. Navy tested its equivalent of air-traffic controllers, it found that operators monitoring multiple computer screens missed a "very high percentage" of changes on those screens. The consequences of such lapses, warned a report, "could be disastrous." See DiVita et al. (2004).

covering, multitasking can be quite dangerous. Take something as simple as talking on your cell phone while driving. In 1999, the U.S. Army studied what effect this has on driving ability. Its conclusion? "All forms of cellular phone usage lead to significant decreases in abilities to respond to highway traffic situations."

This was especially true, the Army noted, for older drivers. Age, it found, plays a significant role in the distracting effect of cellular phone conversations. The older we are, the harder it becomes to screen out distractions. And you don't have to be that old before this ability declines: the dropoff is noticeable after the age of forty.*

> The older we are, the harder it becomes to screen out distractions. And you don't have to be that old: the decline is noticeable after the age of forty.

### Bridge? What Bridge?

Even more worrisome, divided attention can produce a dangerous condition known as inattentional blindness. In this condition it is possible for a person to look directly at something and still not see it. The effect was noted by researchers in the early 1990s; in separate experiments, they found that a surprising number of participants were completely unaware of certain objects presented to them in visual tests. This tendency held true not only when the presented objects were small but when they were large and, presumably, quite obvious.

A real-life demonstration of this effect occurred in 2004 just outside Washington, D.C. On the morning of November 14, James Anthony Jones, a forty-four-year-old charter bus driver, picked up a group of students at the Baltimore/Washington International

---

*Interestingly, the Army's study also found a gender effect: "A point noted without comment is that young females tend to show the least effect of this multi-task performance requirement, while old males clearly show the most effect." See Middlebrooks, Knapp, and Tillman (1999).

Airport for a trip to George Washington's home at Mount Vernon. By all accounts, Jones was in a bad mood that morning. He was upset about the way the driver of the lead bus in the entourage was treating him. The other driver had not only left the airport without him but also failed to talk with him about the details of the trip— in effect, leaving Jones high and dry. Jones called his boss to complain, but apparently got no satisfaction. So he pulled out his cell phone, dialed his sister, and began to vent.

Their route that morning took them along the George Washington Memorial Parkway. The parkway passes through rolling hills and beneath arched overpasses, including a picturesque stone bridge built during the 1930s at Alexandria Avenue. A little more than a quarter of a mile before drivers reach the Alexandria Avenue bridge, there is a large yellow sign on the side of the road. It warns them that the arched overpass ahead has a clearance of just over ten feet in the right-hand lane.

For cars, this is not a problem. But for buses, it is. Jones's bus was twelve feet tall—nearly two feet too tall to fit under this part of the bridge. The cure for this problem is to move toward the center lane, under the peak of the arch, where the clearance is well over thirteen feet. This is what the lead bus did.

But Jones never changed lanes. He never put on the brakes. And he never let off the accelerator. He just continued talking to his sister. Moments later, the bus slammed into the bridge. The collision sheared off the right side of the bus's roof, raining glass down on the students inside and exposing a gaping hole.

"It was surreal," said David Gusella, one of the students on board. "You looked to the right and all you could see was the road. No roof, no windows, nothing."

Incredibly, none of the twenty-seven students on board was killed, though one was seriously injured. After the accident, Jones was interviewed by investigators for the National Transportation

Safety Board, which conducted an inquiry into the accident. His statement is a testament to the power of inattentional blindness. Jones told investigators that not only did he fail to see the yellow warning sign—*he failed to see the bridge.*

## Keep Your Eyes on the Road

Driver distraction is now considered a much more frequent cause of auto accidents than safety officials once believed. In fact, the National Highway Traffic Safety Administration recently revised its estimate of distraction-related accidents. It did this after rigging up cars with cameras and watching what real drivers did as they drove, said Charlie Klauer, one of the researchers involved in the video study.

"We found that basically in about 78 percent of all crashes and 65 percent of near crashes, the drivers were either looking away or engaging in some secondary task," like fiddling with a cell phone or tapping on a BlackBerry, said Klauer. This is much higher than expected. Previous research, which had relied on what drivers *said* they did behind the wheel (as opposed to videotape of what they *actually* did), had indicated that only about 25 percent of crashes were due to driver inattention or distraction.

Klauer said it doesn't take long to distract a driver. A single two-second glance doubles the risk of an accident. So do multiple shorter glances that add up to two seconds or more. That is easily as much time as most of us spend trying to dial a number on our cell phone or entering in a destination on our dashboard GPS unit. One study, in 2004, examined how long it takes drivers to enter an address into a navigation system using a touch-screen keyboard like the ones on leading GPS models. On average, it took a total of eighty-six seconds, or nearly a minute

> It doesn't take much to distract a driver. A two-second glance doubles the risk of an accident.

and a half, to enter the addresses correctly. This typically involved between twenty and thirty-five glances away from the road, depending, in part, on the age of the driver (older drivers needed more glances).

In addition, the task of entering the address proved so absorbing that drivers often veered into another lane of traffic while tapping on the screen—a hazard so common that Japanese regulations prohibit destination entry on the move. But that's not the case in the United States. Here, drivers face few if any restrictions on their use of in-car devices while driving. But this is beginning to change, as the toll caused by distracted drivers mounts. New York City, for instance, is considering a ban on text messaging while driving. The move was prompted by an accident in June 2007 in which five teenage girls riding in a sport-utility vehicle upstate crashed head-on into a tractor trailer, killing all of them. The police later learned from phone records that the driver had been typing messages on her phone just before she swerved out of her lane.

Yet cash-strapped carmakers, eager for extra profits, are increasingly outfitting cars with devices that distract. Onboard navigation systems are now standard on many models. So are entertainment systems for the kids, anticollision warning devices, and rearview cameras. Some carmakers are even installing military-style night-vision systems. BMW, for instance, has begun offering a $1,900 system that uses thermal imaging to spot people, animals, and other obstacles ahead of the vehicle. These images are projected on a small TV screen mounted on the dashboard.

### A Living Room on Wheels—or a Hearse

In 2007, Robert L. Nardelli, the former Home Depot chief executive who now runs Chrysler, told a group of magazine publishers that he thinks cars should be "your most favorite room under your roof."

"I really believe that," he said. "I mean, it has to bring you gratification . . . It's incidental that it gets you from Point A to Point B, right?" Chrysler, which projected a $1.6 billion loss in 2007 from making those point-A-to-point-B cars, has even begun referring to some of its cars as "living rooms on wheels."

And what could be more appropriate in a living room than an entertainment center? As commutes grow longer, Americans are spending more and more time inside their cars; in many areas of the country, the average time behind the wheel is now up to ninety minutes a day. For much of that time (and maybe even all of it), people are bored. And corporations like Microsoft are trying to find a way to entertain them. Speaking at the 2007 Consumer Electronics Show in Las Vegas, Microsoft chairman Bill Gates said the company wants to reach people—no matter where they are or what they happen to be doing.

> "Our ambition is to give you connected experiences 24 hours a day," says Bill Gates.

"Our ambition is to give you connected experiences 24 hours a day," Gates told the crowd. "We admit that, when you're sleeping we haven't quite figured out what we're going to do for you there. But the rest of the time—the minute you get into the kitchen, look at that refrigerator, pick up that phone, hear the alarm clock tell you about the traffic—whatever it is, we want you to have the information that you're interested in. And in thinking about that broadly, one area comes up that clearly demands special work."

That area, he said, is the car.

"Over time," said Gates, "the kind of entertainment you have in the car is going—you're going to want the same great things that you have everywhere else."

To help give people those same great things, Microsoft recently struck a deal with Ford Motor Company. Beginning with its 2008 models, Ford offers a new product it developed with Microsoft. It's called Sync, and it works something like a master brain, allowing you to operate your cell phone, iPod, and other gadgets from one central hub inside the car.

Ford promises that Sync will create "a revolution for in-car communications and entertainment." And the technology is impressive. Using sophisticated speech-recognition software and buttons mounted on the car's steering wheel, Sync allows drivers to multitask in new ways. You can not only play music from a flash drive or an iPod, for instance; you can also create a customized playlist as you are driving by speaking to Sync. And if you happen to get a phone call while Sync

> You can send and receive text messages while you drive. Sync will even translate emoticons, like a smiley face, or well-known texting abbreviations, like LOL.

is playing your music, Sync will display the caller's name on a dashboard screen and pause your music. It will also let you send and receive text messages as you drive. Sync will even translate emoticons, like a smiley face, or well-known texting abbreviations, like LOL.

"It totally integrates, like never before, all of your electronic devices—you know, like your cell phone, Zunes, iPods—all the things that are in your pockets when you get in the car," says Mark Fields, Ford's president for the Americas.

In the past, says Fields, "Sync is the kind of feature that we would have introduced on luxury cars." But its sales potential is so big—"absolutely huge," he says—that Ford will offer Sync as a $395 option on vehicles like the Ford Explorer and Taurus that are aimed straight at the middle class.

## Your Distractible Brain

Ford says Sync is safe because it allows drivers to keep their hands on the wheel.* But when it comes to multitasking, it's not the hands that count; it's the brain. One of the foundations of current knowledge about the effects of multitasking on the brain was laid in 1935. That was when the American psychologist John Ridley Stroop reported that processing the information for one task can cause "interference" with another. Stroop noticed that when study participants were asked to name the color of a word, such as "green," that was printed in an incompatible color—like red—they experienced difficulty saying the color. Now known as the Stroop effect, the phenomenon is thought to occur when two tasks get tangled.

Such a tangling seems more likely as the number and complexity of dashboard devices multiply. The night-vision systems, for instance, require drivers to take their eyes off the road. And many other devices intended to make driving safer work by interrupting drivers at the worst possible time—in the middle of a task. Jim Mateja, an auto critic for the *Chicago Tribune*, discovered this after he took Volvo's $55,000 flagship sedan, the S80, for a spin. Among other things, the car came equipped with a $595 option known as BLIS, for Blind Spot Information System. But BLIS was anything but.

> Many "safety" devices work by interrupting the driver at the worst possible time.

"The system proved only slightly less annoying than fingernails on a chalkboard," concluded Mateja. With BLIS, an orange light in the roof pillar along the windshield flashes to let you know a vehicle has entered the blind spot. But in heavy traffic the light is constantly going off and on, creating a major distraction. And for

*The FAQ section of Sync's Web site (www.syncmyride.com) notes that "this is an especially smart solution as more and more states and jurisdictions outlaw driving while talking on a handheld phone."

a growing number of drivers, that kind of distraction can be a big problem.

When switching from task to task, drivers, just like the office workers we saw at the beginning of the chapter, need downtime to recover. Task switching is especially difficult for older drivers (sixty and up). Their recovery times are much longer than those of younger drivers—sometimes twice as long—and the number of older drivers on the road is growing fast. By 2030, the number of licensed drivers in the United States aged sixty-five and older is expected to nearly double, to fifty-seven million. Not only do older drivers take longer to recover, but their reaction times slow, their visual fields narrow, and they have a harder time processing information.

> Older drivers' recovery times are sometimes twice as long as those of younger drivers.

As a result, safety devices that work by distracting drivers may not work as well as expected. The Air Force found this to be true when it examined CFIT accidents. Many of the aircraft involved in these accidents were equipped with sophisticated electronic devices (not unlike BLIS) intended to warn pilots of impending dangers—such as when they were too close to the ground. But if the devices go off too often, pilots tend to ignore them. And when the alarm is real, the Air Force found, the devices often didn't give the pilots enough time to react. Pilots needed to first figure out what the problem was ("Why is that alarm going off?") and then figure out and execute a solution. One study of information obtained from flight data recorders showed an average pilot reaction time of 5.4 seconds. That may not sound like a lot—but at seven hundred miles an hour it can, quite literally, be a lifetime.

### Coming Soon: A Copilot for Your Car

In response, the Air Force began testing a system that automatically takes over for pilots about to crash into the ground. Automakers are following suit. They are quietly at work developing "workload managers"—essentially, high-tech copilots—that take over the driving for you when you become overwhelmed. The managers work by analyzing information from sensors in the car that monitor such things as speed, braking, and whether the headlights and windshield wipers are being used. In Europe, for instance, certain models of Volvo are equipped with the Intelligent Driver Information System. It will block telephone calls when drivers are changing lanes or turning—situations where drivers should be focusing on the primary task of driving.

> The Intelligent Driver Information System will block telephone calls when drivers are changing lanes or turning.

Meanwhile, the workload behind the wheel continues to increase. Vehicles are not only becoming living rooms, as Bob Nardelli said; they are also becoming offices. This is especially true for the fastest-growing type of vehicle on the road: trucks. Between 1980 and 2003, the number of cars on the road barely budged; but the number of trucks nearly tripled, and they now account for about four of every ten vehicles.

The cabs of 18-wheelers are increasingly filled with distracting gadgets, and their drivers become so absorbed in what they are doing that they frequently rear-end passenger cars in front of them, causing horrific damage. In 2001, for instance, Linda Camacho, a fifty-one-year-old grandmother from Fort Worth, Texas, burned to death after a tractor trailer rear-ended her Buick, causing it to explode. The truck, as it turns out, was owned by Werner Enterprises, of Omaha, Nebraska, one of the largest trucking companies in the world. At the time, the company operated a fleet of nearly eight thousand tractor trailers that logged over one billion miles a year.

After the crash, Camacho's family sued Werner. During the lawsuit her family's lawyers discovered something startling: at the precise moment of the crash the truck's driver was attempting to use e-mail—on a laptop computer.

As incredible as this might seem, e-mailing in trucks is far more common than one might suspect. The e-mail system in the Werner truck was designed by Qualcomm, the San Diego–based wireless company, and was first introduced in the United States in 1988. Since then, its use has grown greatly. According to Qualcomm, its system is used by more than two thousand trucking fleets nationwide. The companies use the system not only to communicate with their truckers by e-mail but to track their positions as well. The messages transmitted across Qualcomm's system in the United States are bounced off satellites and formatted and processed at the company's Network Management Center in San Diego. In 2005, the company estimated, the Network Management Center processed over nine million messages and position reports per day.

Werner could have prevented the drivers of its trucks from e-mailing while driving by sending a lockout signal that would have frozen their computer screens and keyboards while the trucks' wheels were in motion, according to court records. But even after its trucks were involved in a number of serious accidents, including one that killed two people in Pennsylvania, Werner refused to use the lockout signal. A spokesman for the company declined to comment on the case.

*Chapter 6*

# We're in the Wrong Frame of Mind

Not *long ago*, a headline caught my eye:

## MAN MISTAKES PORN DVD
## AS WOMAN'S CRIES FOR HELP
### *He Faces Charges After Entering*
### *Apartment with Sword in Tow*

The man in question is one James Van Iveren, and the story takes place behind Red & Bunny's Diner on South Main Street in the Milwaukee suburb of Oconomowoc. Van Iveren, who was thirty-nine years old at the time, lived behind the diner in an apartment he shared with his mother. On the morning of February 12, 2007, Van Iveren heard sounds—very distinct sounds, he would later say—coming from the apartment above. The sounds were those of a woman screaming.

"She was," he said, "screaming for help."

Van Iveren tried, at first, to ignore the screams. But after a while he could ignore them no longer. He had no phone, so he couldn't call police. So he grabbed the only weapon at hand—a sword acquired as an heirloom—charged up the steps, and kicked in the door of his neighbor's apartment.

There he found his neighbor, Bret Stieghorst, a thirty-three-year-old student at a local technical college, who had been watching a porn film. Its title was *Casa de Culo*. (*Culo* politely translates into English as "rear end.")

"It's all in Spanish, and I don't understand a word of it," Stieghorst told the paper. "I only bought it for the hot chicks."

Van Iveren demanded to know where the woman was. But Stieghorst said there was no woman. He even showed Van Iveren around the apartment, opening closet doors to prove he was hiding no one. Finding no *culo* in the *casa*, Van Iveren left, sword in hand. A little while later the police arrived. They charged Van Iveren with criminal trespass, among other things. Publicity soon followed.

"Now I feel stupid," Van Iveren told reporters. "This was all just a big mistake."

### How We Frame Issues

Stupid, yes. But Van Iveren's misplaced gallantry helps illustrate a very common source of mistakes: framing. A great many day-to-day errors come about because we frame, or look at, an issue in the wrong way. You've probably run into framing problems yourself and didn't even know it: Ever wandered out into the parking lot at the shopping mall and stuck the right key into the wrong car?

Frames work in many ways. Some work through our eyes, but others work through our ears. A few years ago, for instance, researchers in Britain wanted to determine whether music affected the choice of wine bought in grocery stores. So they rigged up a tape deck on the top shelf of a store's wine section. Below the tape deck they displayed four French and four German wines of similar price and dryness. Then

When French music was played, French wine sold well; but when German music was played, sales of French wine plunged.

they played French and German music on alternate days. They found that French wine outsold German wine when French music was played. But when German music was played, the opposite was true: German wine outsold French wine. (This was true even though most of the store's customers generally favored French wines.)

The differences weren't trivial, either. When French music was played, for instance, forty bottles of French wine were sold. But when the German music was played, sales of the French wines plunged to just twelve bottles. The same trend held true for German wines: When German music was played, twenty-two bottles of German wine were sold. But when French music was played, sales fell to just eight bottles of German wine.

Interestingly, most of the customers didn't seem to be aware that they had been influenced by the music. After the shoppers had selected their wines, the researchers asked them to fill out a questionnaire. Of the forty-four shoppers interviewed, only six of them (about 14 percent) said their choice of wine had been influenced by the music. This helps illustrate why frames are so powerful— usually, we don't know they're there.

Much of our understanding of the power of framing comes from the work of the Nobel laureate Daniel Kahneman, a professor at Princeton University, and the late Amos Tversky. Their research focused on how we make decisions, especially under conditions of uncertainty. Through a series of experiments they demonstrated that how we frame an issue can greatly affect our response to it.

In one of their experiments, Kahneman and Tversky divided their subjects into two groups. Both were given the same beginning to a hypothetical problem: the United States is preparing for the outbreak of an unusual Asian disease that is expected to kill six hundred people. Then each group was given a different continuation of the problem and asked which of the options they would favor. Here's the continuation given to the first group:

If program A is adopted, two hundred people will be saved.

If program B is adopted, there is a one-third probability that six hundred people will be saved and a two-thirds probability that no people will be saved.

And here's the continuation given to the second:

If program C is adopted, four hundred people will die. If program D is adopted, there is a one-third probability that nobody will die, and a two-thirds probability that six hundred people will die.

Take a minute to look over the two scenarios (an option not afforded to those participating in the experiment). The two decision problems are the same. Programs A and C, for instance, describe the same outcome: two hundred people will be saved, but four hundred will die. The same is true of programs B and D: there is a one-third chance that all will be saved and a two-thirds chance that no one will be saved.

If people prefer A, then they should also prefer C, because they represent the same pair of consequences. But that's not what happened. In the first group, in which the solution was stated in terms of lives *saved*, 72 percent of the people preferred A. But for the second group, in which the answer was framed in terms of lives *lost*, the preference was reversed: 78 percent voted for D.*

---

*Other experiments have yielded similar findings. In one, people were asked to pick between two treatments for cancer: radiation and surgery. Some were given information regarding *survival* rates; these tended to favor surgery. Others were given information regarding *mortality* rates. A much higher percentage chose radiation. Participants' medical education or experience had little effect on which option they chose; the same outcome was observed among college students, medical students, and doctors.

### Holding On to a Sure Thing

Kahneman and Tversky's findings point to what seems to be a consistent pattern in our decision making. In situations where we expect a loss, we are prone to take risks. When the disease example above, for instance, is framed in terms of deaths, we choose the risky alternative where there is at least *some* prospect of saving everyone. But when we are considering gains, we become more conservative; we simply want to hold on to a sure thing.

This pattern seems to stem in part from the human approach to risk perception.

"There are two systems for analyzing risk: an automatic, intuitive system and a more thoughtful analysis," says Paul Slovic, professor of psychology at the University of Oregon. "Our perception of risk lives largely in our feelings, so most of the time we're operating on system No. 1."

As we'll see later, the system No. 1 approach has significant implications for understanding why we make certain decisions about money. But it also helps explain timid decisions in other, less obvious domains—like football. This was cleverly demonstrated in a recent study by David Romer, a professor of political economy at the University of California, Berkeley. Romer took as his central question the dilemma confronted by every football coach whose team has ever faced a fourth down: Should I do the "risky" thing and go for the first down—or do the "safe" thing and kick?

Using data from over seven hundred National Football League games, Romer began crunching numbers. Along the way, he made certain adjustments to account for the realities of the game. For instance, to avoid the complications that are introduced when one team is well ahead or when the end of a half is approaching, he focused on the first quarter of the game.

When the crunching was done, Romer found that professional football coaches, despite their reputations for toughness, made decisions that were actually quite tame. In particular, he calculated that the percentage of fourth downs in the NFL in which teams are better off "going for it" is 40 percent. But the percentage of downs on which the coaches actually *do* go for it is much lower—only 13 percent.

The numbers themselves are fascinating. But what makes Romer's findings particularly interesting is not so much the data as the coaches. Unlike participants in most academic studies, the coaches are not inexperienced amateurs making wagers for small stakes; instead, they are highly paid professionals (average annual salary: $3 million) who are paid to make precisely these kinds of judgments and whose jobs are on the line (annual turnover among professional coaches is 20 percent). Moreover, the coaches have ample opportunity to learn from their mistakes. Fourth-down situations arise repeatedly in the course of a game, and if a coach makes a bad decision on one fourth-down opportunity, he will probably have a chance to redeem himself later. Not only that, information about other coaches' decisions in the same spot is also readily available.

> The percentage of fourth downs in the NFL in which teams are better off "going for it" is 40 percent. But the percentage of downs on which the coaches actually do go for it is only 13 percent.

And yet despite all this, Romer found that the decisions by these coaches showed a systematic, clear-cut bias that actually ends up hurting their teams. How much it hurts the teams requires some extrapolation, since Romer's study focused on fourth downs that occurred only in the first quarter, and not on the fourth downs that occurred in the remainder of the game. But in all, Romer calculates, going for it more often on fourth downs would correspond to slightly more than one additional win every three seasons. This is a modest effect, but in the hypercompetitive world of professional football not an insignificant one.

The broader implication of Romer's study is obvious: If highly paid professionals are prone to such strong systematic errors in their own decision making, what about you and me? Do we make decisions the same way?

## Framing and Money

One way to answer this is to look at what happens when we face a decision on how to invest our money. Like football coaches assessing a fourth down, most of us don't get out a calculator and tally the risk of various options in mathematical terms. We rely, as Paul Slovic put it, on system No. 1—we want to know how risky an investment *seems*. And that assessment, in turn, often depends on how our potential investment is framed.

Consider the following real-life example: On November 27, 1997, the journal *Nature* published an article reporting positive results for endostatin, a potential cancer cure being developed by EntreMed, a small biotech company based in Rockville, Maryland. On the same day the *New York Times* also ran an article on the drug; the article appeared on page A28. There, for more than five months, the matter rested. Then, on Sunday, May 3, 1998, the *Times* ran a front-page article under the headline "Hope in the Lab: A Special Report; A Cautious Awe Greets Drugs That Eradicate Tumors in Mice." The article favorably mentioned EntreMed and endostatin.

What's important to note here is that the information about the drug didn't change. The article in the Sunday *Times* contained no new news about EntreMed's drug. What did change was the frame: the information went from page A28 to the front page of the Sunday *Times*.

Investors reacted not unlike our friend the swordsman of Oconomowoc: they bounded up the stairs with sword in hand, only to be humbled by what they later discovered. The day after the story appeared in the Sunday *Times*, investors poured hundreds of millions of dollars into EntreMed's stock. In the first two minutes of trading that morning, the price of EntreMed's shares soared by a factor of six. By the end of the day, the stock was still up an astonishing 330 percent. It was, at the time, one of the biggest single-day returns recorded for any stock since 1963. But the euphoria was not to last. Other labs failed to replicate the results initially reported by researchers back in November 1997, and the price of EntreMed's stock collapsed. By October 2008, shares of the company traded at about 34 cents apiece—down considerably from the peak of $85.00 reached the day after the story appeared in the Sunday *Times*.

### How Time Affects Our Decisions

Many factors can affect the way we frame our decisions. One of the least obvious is time. When the consequences of our decisions are far-off, we are prone to take bigger gambles; but when consequences are more immediate, we often become more conservative. A good example comes at the very beginning of life: childbirth. Women's preferences for anesthesia during childbirth have been shown to change with the time horizon. Before the pains of labor set in, many women prefer to avoid anesthesia. But during labor, not surprisingly, they prefer it. Then, a month after the baby is born, they turn against it once again.

Time constraints have been shown to affect our decisions in other ways. After the terrorist attacks of September 11, 2001, for instance, time horizons for many people in the United States shortened. People, especially those in big cities like New York, increasingly adopted a "live for the day" attitude. Activities with long-term benefits, like diet and exercise, were out; treating oneself well in the here and now was in. One result: the diet chain Jenny Craig reported "a huge wave of cancellations."

Timing even affects our choices about the food we eat, the clothes we buy, and the movies we watch. In one experiment, two

> **Those who want to watch a movie later pick highbrow movies; those who want to watch one now pick lowbrow movies.**

groups of people were asked to pick three rental movies, like those available from Blockbuster or Netflix. One group was asked to pick movies to watch later (that is, in the future); the other group was asked to pick the movies they wanted to watch *now*. What happened? The "later" group tended to pick highbrow films like *The Piano*, the Oscar-winning film that portrays the story of a mute woman's rebellion in the recently colonized New Zealand wilderness of Victorian times. But the "now" group picked lowbrow movies like *Clear and Present Danger*, the Tom Clancy action movie starring Harrison Ford.

Researchers found a parallel effect occurred when they asked office workers to choose which of two snacks they'd like to be delivered to them one week later: fruit or junk food. The workers were asked this question at two different times: (1) in the late afternoon, when they were likely to be hungry; and (2) right after lunch, when they were likely to be full. The researchers found that workers do indeed project their current hunger levels onto the future. Some 78 percent of the "hungry" group chose the unhealthy snack; but only 42 percent of the "full" group did. In other words, when the office workers were hungry—which is to say when they wanted their food here and now—they preferred junk food, just as the movie watchers who picked movies to watch immediately preferred trashy movies. But when they were full and content to wait, the office workers preferred healthier food, just as the movie watchers who picked their movies to watch later preferred highbrow movies.

Something similar happens when people buy clothes. Catalog orders for cold-weather gear go up when temperatures go down, which you'd expect—but so do returns, which you wouldn't. Why does this happen? Because when there's a cold snap, people overestimate the use they will make of the items and as a result are more likely to return them. Think of it as the clothes-shopping equivalent of grocery shopping while hungry, says the Michigan State professor Mike Conlin.

> Catalog orders for cold-weather gear go up when the temperature goes down—but so do returns. We realize belatedly that we don't really need that parka.

Conlin and his colleagues studied sales figures for some twelve million items ordered over a five-year period from a large outdoor apparel company. These data included not only the zip code of the buyer but the date of the order and whether the item was returned. Then they matched the sales information with weather records for more than forty-one thousand zip codes in the United States. They found that a plunge in temperature of thirty degrees (to, say, ten degrees from forty) on the order date increases the average return rate of cold-weather gear by nearly 4 percent. Not all items were returned at the same rate. Expensive items like parkas and coats, for instance, tended to be returned about twice as often as hats and mittens. But overall, the increase in returns averaged about 4 percent. That may not sound like much, but catalog orders are a huge business in the United States, with annual sales exceeding $125 billion; even small changes in return rates can have big financial consequences.

### Discovering the Price of Beauty

Many corporations are only now coming to grips with subtle factors like this that can frame their customers' decisions. Not long ago, for instance, a major lender in South Africa began working with Sendhil Mullainathan, a professor of economics at Harvard University. Like all banks, it wanted to make more loans. The question was: How? The classic approach would be to cut the interest rate on the loans to spur demand. But that's not what happened.

Instead, the bank tried an experiment. It mailed letters to more than fifty thousand previous borrowers saying, "Congratulations!" It let them know that they were "now eligible" for a new cash loan. These were small, short-term cash loans somewhat akin to those offered in the United States by "payday" lenders. By American standards, the interest rates being offered were high—between 7.75 percent and 11.75 percent *per month*—though in line with the South African market. And the dollar amount of the loans was small—typically about $150.

Various aspects of the letters were randomized to allow Mullainathan and his colleagues to evaluate the effects of psychological factors mentioned in the letters as opposed to the strictly economic ones, like interest rates. Some customers, for example, were offered a lower rate; some, a higher rate. Others were offered a chance to win a cell phone in a lottery. But the most interesting aspect of the letters, at least for us, appeared in the lower right-hand corner of the letter. Here was a photograph of a bank employee, and the photograph varied by gender and race.

> A woman's photo instead of a man's increased loan demand among men as much as dropping the interest rate five percentage points.

"What we found stunned me," said Mullainathan. "A woman's photo instead of a man's increased demand among men by as much as dropping the interest rate five points!"*

To the more cynical among us, this may not come as a shock—especially after what we learned in earlier chapters about the associations we make with beauty. But to banks, a finding like this

---

*So strong is this effect that in September 2007, the president of the Nigerian Senate said banks must stop using attractive women to persuade customers to open accounts. See Reuters (2007a).

represents a windfall. Even though these were small loans made at high interest, the chance to earn an additional five percentage points in interest with no additional risk represents a rare opportunity.

How transferable are these results from South Africa to other countries?

"It is impossible to tell," says one of the study's coauthors, Marianne Bertrand, a professor at the University of Chicago's Graduate School of Business, because the study concerned only South Africans. But given that the effectiveness of similar psychological manipulations has been documented in the United States, she said, it seems reasonable to think that what worked there could work here.

### Lessons from the Grocery Store

One of those psychological manipulations involves a related effect known as anchoring. Many studies have shown that when faced with a question to decide, people will anchor their responses to almost any number, even implausible ones, especially if that number is given to them first. To test this for yourself, ask a group of friends to write down the first three digits of their phone numbers. Then ask them to estimate the date of Genghis Khan's death. Researchers have done this experiment, along with countless variations (having people estimate the length of the Nile, for instance, or the height of the arch in St. Louis). Time and again, results have shown a correlation between the two numbers. In the case of Genghis Khan, people almost always guess that Khan lived in the first millennium (a date that would have three digits). But he didn't. He lived and died in the second millennium, which has four digits.*

---

*The *Encyclopaedia Britannica* lists three years of birth for the Mongolian warrior: 1155, 1162, or 1167. He died in 1227.

If you knew about this bias ahead of time, of course, you might be able to correct for it. But, just as the existence of frames isn't always apparent (remember the music and the French wine?), neither is the existence of anchors. We often don't realize when information, like prices in stores, is presented in a way that causes us to anchor our purchase decisions one way or another. Yet we are subject to this effect every time we go shopping.

One of the things Vicki McCracken remembers most vividly as a child growing up in southern Indiana is her mother's trips to the grocery store.

"We didn't have lots of money," she told me, "so my mom always looked at prices."

Years later, as a graduate student at Purdue University, McCracken was still curious about those prices. What if shoppers like her mom had a list of prices at the various grocery stores in town *before* they went shopping? Would they change where they shopped or what they bought? They were good questions. So good that the U.S. Department of Agriculture gave her money to find out.

Working with colleagues from Purdue, McCracken devised a simple but extensive test of grocery store prices in four different cities in four different states. In each city they hired price checkers. Their job was to go into local stores with clipboards in hand and write down the prices of certain common shopping items. Then, each week, McCracken and her colleagues published the prices for this market basket of goods in the local newspaper. And not just the total price, either. They printed prices for individual items, like a can of Folgers coffee. The lists brought a quick reaction, but not the kind McCracken expected.

"It got personal," she said. Several stores barred her price collectors from setting foot inside their businesses. One store owner even threatened to kill himself if McCracken didn't stop publishing the price comparisons.

"He had much higher prices than the other stores," she noted.

But McCracken persevered. Rather than have the price checkers walk through the stores with their clipboards—a dead giveaway to managers on the lookout for them—she simply had them go in and buy the groceries whose prices they wanted. (McCracken's group also stopped reporting prices from the store owned by the man who threatened to kill himself.)

The study took a year to complete. Not surprisingly, the researchers found that once the lists started appearing in the newspapers, many of the grocery stores with high prices cut those prices to match those of their lower-priced competitors.

But that wasn't the interesting part.

Not only did McCracken's price collectors track prices on the items they published; they tracked prices on items they *didn't* publish. For instance, they would record the price for a can of Folgers coffee, which they would publish. But they would also record the price for a can of Maxwell House coffee, which they would not publish. This, as it turns out, illuminated a shadowy side of the grocery business.

"As they were lowering the price of the published items, they were increasing the price of the non-published items," McCracken told me. So, as the price of Folgers went down, the price of Maxwell House went up—presumably, she said, to compensate for the discount being offered on the competing brand.

Of course, there is no way most people would know this. And that's the point. Grocers could expect that their customers would anchor their purchase decisions on the sale prices and would flock to the store

> Shoppers tend to anchor their purchase decisions on "sales" prices and flock to the store that advertises them, unaware that they may face higher prices for other goods on their shopping lists.

that advertised them, unaware that they would face higher prices for other goods on their shopping lists.

Similar examples of anchoring abound on the aisles of most grocery stores. Stores wishing to boost sales (and which ones don't?) often use a technique known as "multiple-unit pricing." This is why you will see cans of peaches advertised as "4 for $2" instead of, say, "1 for 50 cents." Logically, the wording of the two advertisements is equivalent (as was the wording of the hypothetical lifesaving problem posed by Kahneman and Tversky); but its effect isn't.

In the first example, the number 4 acts as an anchor. Shoppers see the number and, without really thinking about it, pick up four cans. This influence is surprisingly strong. One field test compared actual consumer purchases in eighty-six different grocery stores. Multiple-unit pricing resulted in a 32 percent increase in sales over single-unit pricing ("1 for 50 cents").

Another way that stores use anchors to get you to buy more stuff is by using quantity limits, such as "Limit, 12 per customer." The number 12 acts as an anchor. Researchers have studied the use of quantity limits to boost sales. And sure enough, they work: the higher the anchor, the higher the sales. Only when the anchor gets absurdly high—say, 50—does the effect tail off.

> A quantity limit—say, 12 per customer— boosts sales. The higher the anchor, the higher the sales.

The key to anchoring is the first number. People tend to process information in the order in which it is presented. And the best place to be in that order is first. Being listed first on the ballot in primary campaigns, for instance, can add up to three percentage points to the results obtained by major party candidates. And simply making the first offer in a negotiation affords a significant advantage. That's because the first offer serves as an anchor for future discussions. This was

> Being listed first on a ballot can add three percentage points to a candidate's results.

empirically demonstrated for the first time in 2001 by researchers in Utah and Germany. They found that whichever party—buyer or seller—made the first offer in a negotiation obtained a better outcome.

### The Power of a Home's Listing Price

This holds true for the biggest negotiation most of us will ever conduct in our lives: buying a house. Houses usually aren't sold in bunches ("4 for $2 million"), and sellers rarely if ever impose quantity restrictions ("Limit, 12 per customer"). But the price we end up paying is nonetheless often anchored to the number we usually encounter first: the listing price.

The power of the listing price was illustrated in a revealing field experiment that pitted college students against seasoned real estate agents. Each was asked to estimate the value of certain properties for sale in Tucson, Arizona. To help them, a ten-page packet loaded with information was given to each of the participants in the study. The information included the kinds of details you and I would probably want to consider when buying a home: recent sales prices in the area, the listing price of homes to be visited, and the standard Multiple Listing Service sheet for each property for sale.

Then they went shopping. The participants were free to look around the house and walk through the neighborhood, just as we would when looking for a new home. When the tour was over, the experts and the amateurs were each offered a calculator and asked to come up with a number of estimates about the value of the home. One of those estimates involved determining the home's advertised selling price. They were also asked to fill out a checklist describing how they had come up with their answers.

The study produced a number of interesting results, probably none of which you will find encouraging if you are trying to buy or sell a home. First, the estimates of the amateurs and the experts weren't that far apart. When it came to the advertised selling price

> **The higher the home's listing price, the higher the estimates of what the home ought to be advertised for.**

of the property, both groups turned in answers that were within a few thousand—and in some cases, a few hundred—dollars of each other. Second, the estimates of both groups were significantly biased by the listing price of the home. Despite all the other information they had, both groups tended to peg their guesses to just one figure: the listing price they were given. The higher the listing price for the home, the higher their own estimates of what the home ought to be advertised for. This helps explain why listing prices may often appear (at least to buyers) to be unrealistically high. The listing price is, essentially, the opening offer in a negotiation: it is the point around which future bargaining is anchored. And, as we saw above, he who makes the first offer gets the better outcome.

Third, and in some ways most important, even the real estate professionals were taken in. Like doctors who take money from drug companies, the real estate agents were blind to their own bias. They thought their professional judgments about a home's value wouldn't be affected by an arbitrarily chosen listing price—but they were. During questioning at the end of the experiment, the real estate agents "flatly denied their use of listing price as a consideration." Yet, when the researchers examined the real estate agents' decision checklists, they found the opposite to be true.*

As the examples above suggest, it's not easy to defeat these psychological manipulations. They are powerful tools, and they work

---

*The researchers said it remained an open question whether the real estate agents really didn't know they had used the listing price in their consideration or whether they were simply unwilling "to acknowledge publicly their dependence on an admittedly inappropriate piece of information." For a similar finding regarding the reluctance of political experts to acknowledge their faults, see Tetlock (1998).

under a variety of circumstances, from grocery stores to voting booths. But here are some tips that may help:

First, try reframing. For instance, if you're in the market for a home, try framing the negotiations not in terms of the price for the whole house ($250,000) but in terms of the price per square foot (say, "I'll offer you $200 a square foot").

Second, be first. This is not always possible, of course. But if a property is in (or about to be in) foreclosure, for instance, it may pay to approach the lender or the owner and be the first to make an offer. That way you put yourself in pole position, and your offer becomes the number around which future negotiations revolve.

Third, be wary of "sales" items. This is often an anchoring mechanism designed to get you to focus on the price that the seller (such as a grocery store) wants you to focus on. But as recent research has shown, the prices of many other items that are not on sale will tend to cancel out the savings from the "sale" item. To check for yourself, save your receipts and compare—or, better yet, try Web sites like www.thegrocerygame.com. It tracks sales prices of thousands of grocery items—both .those that are advertised and those that are unadvertised.

*Chapter 7*

## We Skim

$F$*ew industries make* a habit of confessing their errors. But one does, and on a daily basis, too: newspapers. For anyone interested in mistakes, the correction columns of newspapers often make delicious reading. So delicious that in 2004 Craig Silverman, a freelance writer based in Montreal, launched a Web site, regrettheerror.com, which I heartily recommend. Each year he compiles the industry's greatest hits, as it were, into a book by the same name. It's hard for me to pick a favorite, in part because I have, over the years, compiled so many of my own corrections. Nonetheless, one that springs to mind was published a few years ago in the pages of my alma mater, the *Wall Street Journal.* The text of the entire correction is as follows, with italics added by me:

"Some jesters in a British competition described in a page-one article last Monday ride on *unicycles.* The article incorrectly said they ride on *unicorns.*"

Unicorns don't exist; unicycles do. How, you might ask, could they have missed that?

## We Pay Attention to Beginnings

It's tempting to attribute mistakes like this to simple carelessness. But, as is often the case, the explanation is more complicated than that. When you and I read a newspaper article, the odds are that we don't read every single letter in every single word in every single sentence. We have read enough words and sentences by this point in our lives that we can recognize patterns. If the sentence begins, "The thirsty man licked his...," we know that the final word is probably "lips."

Likewise, if our eyes pick up a short word that begins with "th," we will probably assume that the final letter is *e*, especially if the context is appropriate for the word "the." And indeed, experiments have shown that we do just such a thing. In one, people were asked to read a text and to cross out the letter *e* every time they saw it. It turned out that the later the *e* appeared in the word, the more likely it was to remain undetected. Not only that, the *e* in the word "the" was very likely to be missed—32 percent of the time.

Overlooked mistakes are so common that researchers have given them their own designation: they are called "proofreader's errors." As we will see a bit later, these humdrum errors reveal some interesting quirks about the way human perception works. Perception, above all, is economical; we notice some things and not others. This means that our attention is not always as equally distributed as we might think. Instead, we tend to pay a lot of attention at the beginning of a word (*uni*cycle, *uni*corn), an area that we expect to be rich in cues about what may follow, and less attention later. Investors, interestingly, appear to do the same thing: they pay more attention to financial news released at the beginning of a week, but tend to nod off on Fridays.

> Investors pay more attention to financial news released at the beginning of the week.

Indeed, this tendency is found often enough that it suggests a second, closely related principle: we skim. And the better we are at something, the more likely we are to skim. This is true not only in subjects like reading but in other fields, like music. Musical sight-reading is the ability to play music from a printed score for the first time without the benefit of practice. Good sight readers don't read music note by note; they scan for familiar patterns and for cues to those patterns. Indeed, good sight readers seem to process groups of notes as a single perceptual unit. To use a metaphor: they see constellations instead of individual stars. This is what allows them to play with the speed and fluidity that other musicians must practice to achieve.

## A Mistake Only a Rookie Could Catch

> The better we are at something, the more likely we are to skim. But this ability comes at a cost.

But with this ability comes a trade-off: accuracy is sacrificed, and details are overlooked. This trade-off was documented decades ago by the distinguished piano teacher and sight reader Boris Goldovsky. (Goldovsky, who was best known for his commentary during the Saturday afternoon radio broadcasts of the Metropolitan Opera from 1943 to 1990, died in 2001 at the age of ninety-two.) One day, he discovered a misprint in a much-used edition of a Brahms capriccio—but only after a relatively poor pupil played the printed note at a lesson.

Goldovsky stopped the pupil and told her to fix her mistake. The student looked confused; she said she had played what was written. To Goldovsky's surprise, the girl had indeed played the printed notes correctly—but there was an apparent misprint in the music. At first, the student and the teacher thought this misprint was confined to their edition alone; but further checking revealed that *all* other editions contained the same incorrect note.

Why, wondered Goldovsky, had no one—not the composer, or the publisher, or the proofreader, or scores of pianists—noticed the error? They had all misread the music—and misread it in the same way: they had all inferred a sharp sign in front of the note because in the musical context it *had* to be a G-sharp, not a G-natural.

How could so many experts miss something that was so obvious to a novice? This intrigued Goldovsky. So he conducted an experiment of his own. He told skilled sight readers that there was a misprint somewhere in the piece and asked them to find it. He allowed them to play the piece as many times as they liked and in any way that they liked. Not one musician ever found the error. Only when he told his subjects which bar, or measure, the mistake was in did most of them spot it. (For music fans, the piece is Brahms's opus 76, no. 2, and the mistake occurs in bar 78.)

### A Titanic Error

The world abounds with examples of "Goldovsky" errors overlooked by experts but caught by novices. In April 2008, for instance, a thirteen-year-old schoolboy corrected NASA's estimates on the chances that an asteroid would collide with the earth. A few weeks before that, a fifth-grade boy from Michigan discovered an error at a Smithsonian Institution exhibit that had gone undetected for twenty-seven years. And in 2007, another error was caught, this time by a thirteen-year-old boy in Finland. The incident involved an image of a submarine that the Russian state-owned TV network Rossiya had used to illustrate a story about a Russian submarine voyage to the Arctic. The image, which was distributed by Reuters, was used by news outlets around the world. None of them noticed anything awry. But thirteen-year-old Waltteri Seretin did. The sub, he thought, looked suspiciously familiar. And his suspicions were right: it was a film clip taken from the hit movie *Titanic*, starring Leonardo DiCaprio and Kate Winslet.

"I checked it with my DVD and there it was right there in the beginning of the movie," he told a Finnish newspaper.

## Fool Me Twice

Research since Goldovsky's time has extended his findings. John Sloboda, an internationally known expert on the psychology of music, purposely introduced a number of altered notes into a sample of sheet music. Then he asked experienced musicians to play the music not once but twice. The first time the musicians played the music, Sloboda found, they failed to detect about 38 percent of the altered notes.

But the really interesting thing is what happened when the musicians played the music the *second* time. This time around, the number of proofreading errors didn't go down—it went up! This suggests that the musicians had become familiar enough with the music after just one playing that they were no longer playing it note by note, but by looking for patterns. In short, they were skimming.

This tendency has profound implications for understanding why we don't detect many of our errors: as something becomes familiar, we tend to notice less, not more. We see things not as they are but as (we assume) they ought to be. This ingrained behavior can cause us to overlook not only small things, like altered musical notes, but some that are startlingly large.

> As something becomes more familiar, we tend to notice less, not more. We come to see things not as they are but as (we assume) they ought to be.

A case in point occurred a few days before the Halloween of 2005, in the small town of Frederica, Delaware. There, the apparent suicide of a woman found hanging from a tree went unreported for more than twelve hours—even though her body was plainly visible for much of that time to neighbors and passersby. At about 9:00 p.m. the previous night the forty-two-year-old woman apparently climbed the tree and then used a rope to hang herself, directly across the street from homes on a moderately busy road. At the time, of course, it was dark. But come daylight, her body, suspended about fifteen feet above the ground, could easily be seen from passing vehicles. Yet no one called police until nearly eleven in the morning—some fourteen hours after the woman had hanged herself.

"They thought it was a Halloween decoration," explained the wife of the town's mayor, who was among a small crowd of people standing across the street when police arrived.

## The Importance of Context

As sad as this example is, it illustrates how much we rely on context to guide our perception of everyday events. Context is the great crutch; we lean on it much more than we know. If something happens around Halloween, we assume (perhaps without really thinking about it) that it's Halloween related. And most of the time—who knows—it probably is. But when it isn't, the odds are high that we will overlook it, as the neighbors did with the hanging.

Encountering something or someone out of context makes recognition far more difficult; it becomes much harder to place a face. Maybe you've run into somebody you can't quite place. Is it the guy from the dry cleaners? Or maybe someone from your kid's karate class? Until you know *where* a person belongs, you often can't recall *who* he is: what you're missing is the context.

This is true not only for things we see (like a body hanging from a tree) but for things we read. Reading, at first blush, may strike you as the most literal of tasks: the words are printed in black and white and confined within the four corners of the page. But to understand certain prose passages, context is a prerequisite. Consider the following passage:

The procedure is actually quite simple. First, you arrange things into different groups depending on their makeup. Of course, one pile may be sufficient depending on how much there is to do. If you have to go somewhere else due to lack of facilities that is the next step, otherwise you are pretty well set. It is important not to overdo any particular endeavor. That is, it is better to do too few things at once than too many. In the short run, this may not seem important, but complications from doing too many can easily arise. A mistake can be expensive as well. The manipulation of the appropriate mechanisms should be self-explanatory, and we need not dwell on it here. At first, the whole procedure will seem complicated. Soon, however, it will become just another facet of life. It is difficult to foresee any end to the necessity for this task in the immediate future, but then one never can tell.

What's it all about, Alfie?

The answer: washing clothes.

Ah, it makes sense now, right?

This is what the researchers John Bransford and Marcia Johnson discovered. They gave this passage, and others like it, to a number of people. They then asked their subjects to read the passages and tell them what the passages were about. They found that contexts are important for processing incoming information. Without context, many of the subjects in their reading experiments were stumped. They had no idea what was being discussed in the passage. Although they searched for a situation the passage might be about, they were generally unable to find one suitable for understanding the entire passage, though they could make parts of it make sense. But if they were given the context ahead of time, presto, it all made sense. This is one of the reasons why newspaper articles have headlines and photographs have captions—they provide context, allowing us to quickly understand what it is we're looking at.

## How a Walk in the Park Improves Memory

Context is also important when it comes to remembering things. When something is out of context, it is not only harder to recognize, it is harder to remember. But reinstate the context, and memory improves. This was demonstrated years ago in a deceptively simple experiment with preschoolers who were taken for a walk in the park. The day after the walk, the children were asked to recall what they could of their visit to the park. The results showed that when the children were asked to make the recollections while they were sitting in a quiet room, their recall was relatively poor. But when the children were taken back to the park, they recalled significantly more activities.

> **When children were taken back to a park, they remembered significantly more about a prior visit.**

Adults show the same tendency. In a well-known experiment performed more than thirty years ago, British researchers asked people to memorize a list of words. Some of the people were asked to memorize the words while on land; others were asked to memorize the words while underwater. (This was not without risks; those underwater were equipped with scuba gear, and one diver was nearly run over by an amphibious military vehicle.) Then the researchers tested the recall of their subjects.

They found that what was learned underwater was best recalled underwater, and what was learned on land was best recalled on land. On average, for instance, the on-land learners remembered 13.5 words when they were tested on dry land; but when they were tested underwater, they remembered only 8.6. The same trend held true for the divers: when they were tested underwater, they remembered an average of 11.4 words; but back on dry land, they remembered only 8.4 words.

### Happy Endings

These findings have been extended even further to include not only physical contexts but emotional ones as well. Events learned in one emotional state are best remembered when we are back in that state. Happy times, for instance, are best remembered when we're happy.

In one experiment, people were made to feel happy or sad by a posthypnotic suggestion they were given as they read a brief story about two college men getting together and playing a friendly game of tennis. One of the men—Andre—is happy; everything is going well for him. The other—Jack—is sad; nothing is going well for him. After the hypnotized people finished reading the story, they were asked to tell the researcher who they thought was the central character and which one they identified with. The researchers found that the happy readers identified with the happy character, thought the story was about him, and thought the story contained more statements about him; sad readers, on the other hand, thought the opposite: they identified with the sad character and thought there were more statements about him.

The moral of these stories is that context matters. The location of the context may vary: sometimes, as with the Halloween suicide, it is in the world; but other times, as with happiness, it is in our heads.

*Chapter 8*

# We Like Things Tidy

H*ere's a good* bar bet: Ask the guy next to you to indicate the direction between the following pair of cities: San Diego, California, and Reno, Nevada. Most people will say, correctly, that Reno is north of San Diego. Then ask whether Reno is east or west of San Diego. Most people will say that Reno is east of San Diego. But it's not. It's *west* of San Diego. (Go ahead, check a map.)

Why do so many people get this wrong?

One part of the answer involves how we remember maps. And by maps I don't mean just paper maps like those in an atlas; I mean the maps we carry in our heads of real-world places like our neighborhoods or the parking lot at the local mall, or the path to our favorite fishing hole. The problem is that in remembering maps, we systematically distort them. We straighten curved lines, make odd shapes more symmetrical, and align parts that shouldn't be aligned. In short, we clean up the picture.

> The problem is that in remembering maps, we systematically distort them.

This tendency was demonstrated years ago by the late social

psychologist Stanley Milgram. Milgram researched wide-ranging topics like the lost-letter technique, in which people encounter an unmailed letter lying on the ground (the interesting part is what they do with it), and the small-world problem (the famous "six degrees of separation"). But he is best remembered for a series of experiments begun at Yale University in the early 1960s that demonstrated a less-than-admirable human tendency: blind obedience to authority. In these experiments, volunteers were induced to give apparently harmful electric shocks to an actor posing as a pitifully protesting victim. As the actor gave wrong answers to certain questions, the volunteers were told to give shocks of greater and greater intensity. At 120 volts, the actor would shout that the volts were becoming too painful. At 150 volts, he would demand that the experiment end. And so it would go until at the higher levels there came only silence from the cubicle in which the actor sat. The volunteers could have quit giving shocks at any time. But most did not. A full 65 percent of Milgram's subjects—both men and women—continued applying shocks until the very end, simply because they were commanded to do so by a figure of authority.

## We Straighten the Seine

Milgram's experiment with maps demonstrated a different—and much more benign—human trait. Milgram simply asked residents to draw maps of Paris. He and a colleague collected hundreds of hand-drawn maps from architects and butchers, the young and the old, college educated and not. The maps, as you might expect, were quite varied. Some omitted famous places like the Eiffel Tower and Notre Dame; others contained minute descriptions of obscure one-way streets. But when Milgram examined the maps closely, he noticed something striking: residents consistently straightened the Seine, the river that flows through the heart of the city. In all, 92 percent of the people asked to draw maps of the city understated the curvature of the Seine.

This tendency is not unique to Parisians. Cabdrivers were found to do the same thing when they were asked to draw the streets of New York City: they straightened them. Not only that, researchers have found that numerous other errors creep in when people are asked to re-create maps from memory. Distances, for instance, are often fouled up. Short distances are overestimated, and long distances are underestimated. When people are asked to navigate by using landmarks, such as their homes or a famous building nearby, something even stranger happens: they judge the distance *to* a landmark to be less than the distance *from* a landmark. This holds true even on large-scale comparisons. For instance, people judge North Korea as being closer to China than China is to North Korea.

Our propensity to distort maps has been studied for decades by Barbara Tversky, who recently retired from her position as a professor of psychology at Stanford University and now teaches at Columbia University's Teachers College. In one experiment she presented students with two maps: an altered one, showing North America as being more or less directly above South America; and an accurate one, depicting South America as being to the southeast of North America. Most students picked the inaccurate one, indicating that their mental representations of the world are distorted toward a simpler organization. Moreover, these errors weren't confined to the Americas. People systematically misalign other parts of the globe as well; they erroneously think that Rome, for instance, is south of Philadelphia.

> **Most people think—erroneously—that Rome is south of Philadelphia.**

Given the kinds of distortions found by Tversky, it is reasonable to wonder how people are able to get from point A to point B.

Not like other creatures, it seems. Take one of nature's true Daniel Boones, for instance: honeybees. Like other insects, bees navigate by dead reckoning: they keep track of how far they've gone and in which direction (an impressive feat, given that bees can fly more than six miles to find food). Then, when it's time to return, they are able to figure out the shortest way back home: they make a beeline.

## An Information Hierarchy

But people, by and large, don't make beelines. Experiments by Tversky and others indicate that another way the human mind responds to all the information it needs to form a map is by organizing the information into a hierarchy. This is what happens when we look up at the stars. The positions of stars have traditionally been remembered by aligning them into meaningful structures, such as constellations. It becomes much easier to find the North Star, for instance, by first locating the Little Dipper, then following the Dipper's handle to the end.

Down on earth, we often use a similar tactic. Instead of remembering the exact positions of innumerable cities, we remember the positions of larger geographic units, like states. Then we remember which cities belong to which states and use the locations of the states to guide us to the locations of the cities. The states, in effect, work like constellations, organizing cities into useful units.

But this approach can result in systematic distortions of its own, which brings us back to the bar bet that began this chapter. Most people think Reno is east of San Diego because they reason as follows:

1. San Diego is in California, and California is on the West
   Coast;

124°    120°    116°

NEVADA

— 40°

• Reno

CALIFORNIA

— 36°

West

San
Diego

Reno 119.85°
San
Diego 117.13°

— 32°

2. Reno is in Nevada, and Nevada is east of California;

3. Therefore, Reno must be east of San Diego.

But it's not.

As you can see from the map above, the southern tip of California actually curves in quite a bit to the east (note the longitude lines: San Diego is at the 117th meridian west; Reno is at the 119th). But in remembering California's shape, we tend to straighten this curve, just as Parisians straighten the Seine.

The tendency to reshape the irregular features of our world into smoother, more symmetrical forms is not limited to the things we see, like the rivers of Paris or the streets of New York. We show the same inclination when it comes to things we read and even things we hear. Inconvenient details tend to be pruned from our memories, and facts that do not fit together in a coherent way tend to be forgotten, de-emphasized, or reinterpreted.

## "The War of the Ghosts"

Perhaps the most famous example of this particular bias comes from an old Native American folktale, "The War of the Ghosts." The story dates from the late nineteenth century and was originally told in the Kathlamet language, a dialect spoken by Chinook people who lived along the Columbia River, between modern-day Washington and Oregon. By the time the story was translated into English and published, in 1901, the language was nearly extinct; a researcher from the U.S. Bureau of American Ethnology was able to find only three people who spoke it. The story survives today largely because it was read, half a world away, by a psychologist at Cambridge University, Sir F. C. Bartlett.

Bartlett was an anomaly at Cambridge. The son of a boot maker, he was educated at home in his teens because his parents thought his health was too poor, following a bout of pleurisy, for him to go away to school. His illness, though, had a silver lining: it allowed him to read widely and to spend time walking the countryside, where he developed a tendency to observe everyday human activity. This penchant for observation had a profound effect on Bartlett's professional life. In contrast to that of his peers, Bartlett's approach to psychology would rely heavily on observation, rather than on academic thinking.

In the early part of the twentieth century, Bartlett engaged in a lengthy series of experiments on perception and memory. He believed that we do not passively notice the world around us; instead, our awareness of events is both selective and constructive. Bartlett thought that much of what strikes the eye is never seen or remembered. Conversely, he also believed that much of what we *think* we have perceived or recalled never, in fact, occurred. Moreover, he believed that recall could be affected by a number of factors, among them culture. Bartlett was able to show, for instance, cases in which Africans had selectively remembered events of little importance to Britons, or vice versa.

Enter "The War of the Ghosts." In one of his experiments, Bartlett gave the story to twenty Englishmen, seven of them women and the rest men. He asked them to read it and then, over a period of time, to write down as much of it as they could recall. What, he wanted to know, would a twentieth-century Englishman make of a tale told by a nineteenth-century American Indian? Since the same question applies equally well to those of us in the twenty-first century, take a few minutes yourself to read the story. If you want to test yourself against Bartlett's subjects, put this book down after you've read the story and take a break. Then take out paper and pen and write down as much of the story as you can remember. Without rereading the story, do the same thing a week from now and compare the two versions.

Here's the version Bartlett gave to his subjects:

### The War of the Ghosts*

One night two young men from Egulac went down to the river to hunt seals, and while they were there it became foggy and calm. Then they heard war-cries, and they thought: "Maybe this is a war-party." They escaped to the shore, and hid behind a log. Now canoes came up, and they heard the noise of the paddles, and saw one canoe coming up to them. There were five men in the canoe, and they said:

*The text of the original translation has been digitized as part of the Internet Archive book project. It's freely available at www.openlibrary.org/details/no26bulletinethn00smituoft.

"What do you think? We wish to take you along. We are going up the river to make war on the people."

One of the young men said: "I have no arrows."

"Arrows are in the canoe," they said.

"I will not go along. I might be killed. My relatives do not know where I have gone. But you," he said, turning to the other, "may go with them."

So one of the young men went, but the other returned home.

And the warriors went on up the river to a town on the other side of Kalama. The people came down to the water, and they began to fight, and many were killed. But presently the young man heard one of the warriors say: "Quick, let us go home: that Indian has been hit." Now he thought: "Oh, they are ghosts." He did not feel sick, but they said he had been shot.

So the canoes went back to Egulac, and the young man went ashore to his house, and made a fire. And he told everybody and said: "Behold I accompanied the ghosts, and we went to fight. Many of our fellows were killed, and many of those who attacked us were killed. They said I was hit, and I did not feel sick."

He told it all, and then he became quiet. When the sun rose he fell down. Something black came out of his mouth. His face became contorted. The people jumped up and cried.

He was dead.

Bartlett found that when his subjects attempted to recall the story, it underwent significant changes. First, it was invariably shortened. On the first attempt at retelling,

> On the first attempt at retelling, participants usually cut the length of the story in half.

participants usually cut this length in half. Second, details were cut, changed, or even made up. The weather ("foggy and calm") disappeared from many of the accounts, as did the seals. Forests appeared where there had been none before. In some retellings, the men became brothers. In others, even the title of the story was changed. Third, the language was altered in subtle but significant ways. Odd words were turned into more conventional ones, and the tone generally became more conversational.

Finally, and perhaps most important, the story was rationalized. Bartlett knew that a mystical nineteenth-century Native American story like this one might confound cerebral twentieth-century residents of Cambridge, England. That's one of the reasons he picked it. He wanted to know how his "educated and rather sophisticated subjects" would deal with a story drawn from a "culture and a social environment exceedingly different from those of my subjects."

The answer, in short, is that they didn't deal with it all that well.

"Hardly ever, at the outset, was there an attitude of simple acceptance," Bartlett would note. Instead, readers tried to fit the story into their own existing ways of understanding the world. For instance, after reading the story, one of the readers confidently declared, "This is very clearly a murder concealment dream." Maybe it was, and maybe it wasn't. But to Bartlett, one thing was clear: once this type of "fitting" had been made, the form of the story was irreparably changed. "Before long," Bartlett found, "the story tends to be robbed of all its surprising, jerky and apparently inconsequential form and reduced to an orderly narration." In other words, the story had been simplified, smoothed, and straightened.

We can see some of these same processes at work in our everyday lives. Say you're walking down the aisles of your local Costco, glancing at the prices as you push your cart along. Which ones do you remember—and why? It turns out that when we try to recall prices, we exhibit a "shortening" effect akin to the one that Bartlett's subjects exhibited when they tried to recall their story. In particular, researchers found that each extra syllable in a price reduces the chance of its being recalled by 20 percent. This is true even if the prices have the same number of digits. So a price of, say, $77.51 (which has eight syllables) is less memorable than a price of $62.30 (which has five syllables). The smoother, rounded price sticks with us—as do memories of smoother, rounded maps and stories.

## O Say Can You Remember the Words to the National Anthem?

A key point for Bartlett was that the thing being remembered often reflected the nature of the person doing the remembering. Different people exposed to the same scene may remember that scene differently—not necessarily because they have better or worse memories, but because of *who* they are. As Bartlett put it, the whole rationalization process "tends to possess characteristics peculiar to the work of the individual who effects it and due directly to his particular temperament and character."

Remembering something verbatim, it turns out, is very hard to do—even if it's something you've recited hundreds of times since grade school and think you know by heart. Take "The Star-Spangled Banner," for instance. It's not a very long song. It takes a little over a minute to sing and has just eighty-one words. But how many of them do you know? If you want to have some fun, get that pen and piece of paper back out and write down as much of the lyrics as you can remember—without singing the song in your head.* Then check your version against the original version by Francis Scott Key, which is reprinted at the end of this chapter.

---

*Or go to Idolator.com and visit the "Hey, it's Enrico Pallazzo [*sic*]!" Hall of Fame, named after the opera singer impersonated by Leslie Nielsen in the screwball comedy *The Naked Gun*. Inductees include Michael Bolton, who muffs the words to the national anthem while singing at a 2003 American League championship game—and peeks at his palm, where he'd apparently written down the lyrics.

David Rubin, a professor at Duke University, once asked a group of undergraduates to do the same thing. He found that they remembered, on average, just thirty-two words of our national anthem.

But what happens to that recall when the words are put to music? Rubin found that when he played the instrumental music to "The Star-Spangled Banner" first and *then* asked people to write down the lyrics, recall jumped from thirty-two words to fifty-two words. Not perfect, but not bad. Rubin and a colleague, Wanda Wallace, have concluded that key aspects of music, like its melody, act as constraints against forgetting. In addition to rhythmical information, melody can provide clues to the singer about the length of the lines and the intonation patterns of the words that help us remember the text.

> After instrumental music was played, recall of the national anthem's lyrics jumped from thirty-two words to fifty-two words out of eighty-one—not perfect, but not bad.

### How Much of What You Say Is True?

The process of retelling a story in our own narrative style acts in somewhat the same way as a musical setting. It places certain constraints on what we recall, and these constraints guide our reconstruction of events. If we tell a story in a funny way, for instance, we're likely to leave out certain details or maybe even add a few of our own making. In the process, the story doesn't simply become *a* version of the original event, it becomes the event; it is the way we remember it. Likewise, when people ask us how our day was, what do we tell them? We can't tell them everything— Good Lord, they'd be bored to tears. So how do we decide what to

> If we tell a story in a funny way, the story doesn't simply become *a* version of the original event—it becomes the event; it is the way we remember it.

leave in and what to leave out? In short, how do we organize the universe of facts available to us?

One answer comes from Barbara Tversky, the professor who has studied how we misremember maps. Her research indicates that we tend to organize the narrative events in our lives much the same way as we organize the stars in the sky: we use a hierarchy to shape a constellation from the many facts confronting us.

"Things don't happen in words," she said. "They happen in all kinds of senses at the same time. If you're witnessing an accident, nobody is narrating it while it is happening, and you are paying attention to certain things and not other things." As a consequence, she said, we are forced to go back and impose an order on the story we will tell.

A few years ago she and her colleague Elizabeth Marsh of Duke University asked students to keep daily records over a period of several weeks of the stories they told. In particular, they asked the students to record the gist of their stories, as well as the audience they told them to (parents or other students, for instance) and the purpose of the story (to inform, to entertain, and so on). The students were also asked to note whether they had distorted the story in any way—by embellishing certain parts, for instance, or leaving out others. Finally, the students were asked whether they had misrepresented any events.

After tallying the results, Tversky and Marsh found that most of the stories—58 percent—were told to inform. They typically involved social events and were retold, on average, 2.7 times. None of this was unexpected. What was unexpected, though, was the degree of distortion. Not only did the students admit to bending the truth quite a bit; they bent it even more than they thought they had. Tversky and Marsh found that the students had added, omitted, exaggerated, or minimized information in 61 percent of

> Students added,
> omitted,
> exaggerated, or
> minimized
> information in 61
> percent of their
> stories.

their stories. Yet, when the students were directly asked how often they had done so, they admitted doing so only 42 percent of the time. That's a significant difference, and it suggests that some of the distortions were so common that students didn't think of them as lies.

The most common alteration was omitting important details, which was reported in 36 percent of the stories. Exaggeration and minimization occurred about equally, appearing in 26 percent and 25 percent of the stories. And 13 percent of the stories contained outright fabrication—information that was not part of the original event. Moreover, students tailored their stories not only to their audience but, importantly, to their purpose. With stories told to convey information, for instance, students tended not to exaggerate—though they did tend to minimize and omit important details. With stories told to entertain, on the other hand, the students tended to do the opposite: they would exaggerate and add details, but not minimize or omit important information.

Some of the results found by Tversky and Marsh are strikingly similar to results obtained in a separate study. In that study, people were recorded while interacting with a new person for ten minutes. Afterward, the tapes were played back, and the people were asked to identify how much of what they said was truthful; 60 percent of them admitted to lying during the conversation.

### Lying—or "Impression Management"?

What could account for so much fibbing?

Part of the explanation, Tversky believes, lies in the assumptions we make about the purpose of the stories we hear.

"We have this Anglo-Saxon idea that talk is about information," Tversky told me one autumn afternoon as she strolled through New York's Central Park. But it's not—at least not all of the time. Instead, she said, think of conversation not as a means of truth telling but as a form of behavior designed to achieve a particular end.

> We think conversation is about imparting information—but it's not. Sometimes it's a form of impression management.

"If you think about talk as a behavior, we behave in ways that will make people think certain things or act in certain ways toward me—to like me, or to think that I'm a smart person or a strong person or whatever."

In this sense, she said, the purpose of conversation isn't to convey the truth—it's to create an impression. So accuracy tends to take a backseat to impression management.

Given that the stories told by the students contained so much distortion, it's reasonable to wonder whether the people listening to them could tell they were being misled. Unpublished research by Tversky and Danny Oppenheimer of Princeton suggests that they could, though not all the time. They found that listeners missed some distortions that did occur and likely "caught" others that didn't occur. In particular, listeners reported far more exaggerations than the storytellers admitted including. What tipped them off? Often, it wasn't the story but the teller. "She's a drama queen," one of the listeners said about one of the storytellers. "He was waving his arms too much," said another. Said a third: "He wouldn't tell me about the drugs with his parents in the room." Whatever the cue, the listeners were not as gullible or as credulous as we might think; the discount they applied to the stories was pretty much on the mark. And men, interestingly, were far more skeptical of what they were hearing than women were.

"By a long shot," says Tversky.

But the most interesting effect of all was not on the listeners but on the tellers. In retelling their stories, Marsh and Tversky found, storytellers misled not only others but themselves. The researchers found that the alterations introduced during the storytelling phase became incorporated into the memory of the storyteller—so much so that the storytellers often "remembered" things that weren't true. This is hard to demonstrate in the real world, but a bit easier to do in the lab, where conditions can be more closely controlled. So students were given fictional stories about two new roommates, who, for simplification purposes, we'll call Michael and David. Each had admirable traits. But each also committed annoying acts like spilling red wine on the carpet or borrowing a jacket without asking.

> **Men were more skeptical than women about the stories they heard—"by a long shot."**

Later, the students were asked to write a letter about their roommate. In some cases, the letter was to be favorable—recommending the roommate for membership in a fraternity. In other cases, the letter was negative—written to the university housing office in an attempt to get the roommate kicked out. After the letters were written, the students were asked to recall as much as they could about the original stories they had been given. But what they recalled was often wrong. Some of those who had written a letter to the housing office about Michael remembered that he had spilled wine on the carpet. But he hadn't; David had.

In effect, we really do come to believe our own untruths. But this process, just as Bartlett had found, is unwitting.

How many of the eighty-one words did you get correct?

### The Star-Spangled Banner

*O say can you see by the dawn's early light*
*What so proudly we hail'd at the twilight's last gleaming*
*Whose broad stripes & bright stars through the perilous fight*
*O'er the ramparts we watch'd, were so gallantly streaming?*
*And the rocket's red glare, the bomb bursting in air,*
*Gave proof through the night that our flag was still there*
*O say does that star-spangled banner yet wave*
*O'er the land of the free & the home of the brave?*

*Chapter 9*

## Men Shoot First

Aspects of our personalities predispose many of us—and maybe even all of us—toward certain kinds of errors. If you like to drive fast, for instance, the odds are that you also like to trade stocks a lot. In one study of such drivers in Finland, each additional speeding ticket the drivers received was correlated with an 11 percent increase in the turnover rate of the driver's stock portfolio. In fact, if you get more than your share of speeding tickets, the odds are that you not only trade a lot but trade too much. This is where the error comes in. Several studies have demonstrated that those investors who trade the most typically earn the least. One study during the bull market of the 1990s calculated the stock market's average annual return during this period at 17.9 percent. But those who traded the most during this period earned far less—just 11.4 percent.*

> **Investors who trade the most typically earn the least.**

Interestingly, the researchers from the Finland study found that

---

*If you're wondering how you stack up, here's a benchmark you can use: on average, investors turned over 75 percent of their stock portfolios each year; in other words, they sold three out of every four stocks they held.

excessive trading was related not only to the number of speeding tickets the driver had received but to something else: the driver's gender. As a general matter, men tend to rack up more speeding tickets than women do. This is true for young men and old, though the gap diminishes with age. Men also tend to trade stocks more often than women do—45 percent more often, according to one study. This gap, too, persists throughout life, though it does wane a bit for older men. And when the men aren't married, the difference is even starker: single men trade 67 percent more often than single women do.

### Overconfident Men

*Why* they do this is open to debate. A leading explanation is differing confidence levels between men and women. As we'll see in the next chapter, overconfidence is a leading source of human error, and across a wide field of endeavors both men and women have been shown to be overconfident.

But men, as a rule, tend to be more overconfident than women are, and this difference explains much about the kinds of mistakes men and women make. When men and women are asked to estimate their own IQs, for instance, men, on average, will give higher estimates than women will. But men aren't as smart as they think they are; their IQs turn out to be lower than they had guessed. Women, on the other hand, turn out to be smarter than they think they are: their IQs are, on average, *higher* than their estimates. In other words, men overestimate their IQs, and women underestimate theirs. And men—but not women—have also been shown to overestimate their own attractiveness.

> Men tend to overestimate their intelligence—and their attractiveness.

This confidence gap is especially pronounced when it comes to male-dominated fields like war and finance. Men, even as boys, think they are better at these kinds of things than women do.

School-age girls, for instance, have been shown to underestimate their grades in the traditionally male-dominated field of mathematics—but boys don't. Men also tend to be more trigger-happy than women are. This is what the U.S. Army found when it studied simulated incidents of friendly fire—that is, U.S. troops shooting one of their own. When the Army examined the results, it found that male soldiers tended to shoot people they shouldn't shoot, but female soldiers did the opposite: they tended *not* to shoot people they *should* shoot. In other words, the men shot the good guys, and the women failed to shoot the bad guys.

A separate study found that men were also much more inclined to start a shooting war. The Princeton professor Dominic Johnson and half a dozen colleagues from Harvard and the University of California recently ran a series of experimental war games, not unlike those used by the Pentagon to prepare for the real thing. Among other things, Johnson and company wanted to examine the relationship between gender, aggression, and overconfidence. So they gathered nearly two hundred men and women, sat them at computer terminals, and asked them to assume the role of a leader of a country involved in a conflict with another country over diamond deposits discovered along a disputed border.

The volunteers competed against each other in pairs and were not aware of the identity or sex of their opponent. They played six seven-minute rounds, and in each round they had to make a decision about what sort of action to take. They could do nothing, they could negotiate, they could start a war, or, if they felt like it, they could even surrender (though no one ever did). Before the games began, the researchers asked each player to predict how well he or she would do. If they thought they were the best, they were to rank themselves No. 1; if they thought they were the worst, they were to rank themselves No. 200; if they thought they'd end up in the middle, they were to rank themselves No. 100. After the games were over (but before the results were known), the researchers again asked them to forecast their rank.

As you might expect, the volunteers, on average, thought they would do better than they did. But when the researchers looked at these self-rankings more closely, they found that the overconfidence was entirely attributable to the self-ranking of the men; the self-rankings of the women in the study were all about 100, which is to say right in the middle.

More important, Johnson and his colleagues found that men made significantly more unprovoked attacks than women did. Why? Overconfidence. After analyzing their information, the researchers found that men were overconfident about their odds of success in war, and that men who were more overconfident were more likely to launch wars.

> Men were overconfident about their odds of success in war, and the men who were more overconfident were more likely to launch wars.

The experiment took place in a lab, under admittedly contrived conditions that one wouldn't necessarily find in the real world. Still, it is almost impossible not to see parallels to war making in the real world. Remember when George Bush asked the then CIA director, George Tenet, how confident he was that Saddam Hussein possessed weapons of mass destruction?

"Don't worry," Tenet reportedly replied. "It's a slam dunk."

### Risk and Reward

Just because men *think* they are better at some things, of course, doesn't necessarily mean that they *are* better. Often, their performance is worse. When it comes to stocks, for instance, all that trading by men reduced their net returns by 2.65 percentage points per year, versus just 1.72 percentage points per year for women. This yields a small but still significant difference of 0.93

percentage points per year. And that is for *all* men. If you look only at the trading records of unmarried men (whose decisions, presumably, are not subject to influence by a wife), their performance versus that of women is even worse: a gap of 1.44 percentage points per year.*

Many studies over the years have shown that men and women perceive and remember aspects of their lives in different ways—often from a very young age—and that the roots of some of our mistakes can be traced back, at least in part, to these differences in perception and memory. Take, for instance, the way men and women perceive risk. Across a variety of areas, women have been shown to be more risk averse than men—a finding that appears to be reflected in the Army's friendly-fire study. When the female soldiers were confronted with a risky situation—shoot or don't shoot?—they typically chose the more risk-averse option: don't shoot.

This same tendency to avoid risk can be seen in everyday activities, like driving a car. Women, for instance, report wearing their seat belts more often than men do. Men, on the other hand, are prone to engage in risky behavior—like running yellow lights—more often than women are. Given such differences, it is not surprising that men in the United States are three times as likely as women to be involved in fatal automobile accidents. They are also more likely than women to die from drowning or accidental poisoning.

> **Men are three times as likely as women to be involved in fatal automobile accidents.**

But there is more to this risk avoidance than meets the eye. The Columbia University professor Elke Weber and her colleagues

---

*It's worth noting that when it comes to picking stocks, both men and women tend to be lousy. The stocks they choose to sell reliably earn greater returns than the stocks they choose to buy.

have studied how men and women perceive different kinds of risks. In particular, they have focused on five types of risks:

1. Financial
2. Health and safety
3. Recreational
4. Ethics
5. Social

A few years ago they gave questionnaires to more than five hundred men and women, from teenagers to people in their mid-forties. For each category of risk, the people in the study were asked roughly twenty questions. For instance, in the "recreational" category they might be asked if they would try bungee jumping; in the "financial" category they might be asked whether they would cosign a car loan for a friend; and in the "social" category they might be asked whether they would speak their mind on an unpopular topic at a social occasion. They were asked to answer each question by assigning it a risk rating on a scale from 1 to 5, with 1 being "not at all risky" and 5 being "extremely risky."

In four of the five areas examined, Weber found that women appeared to be significantly more risk averse than men. (The one exception was the area of social risk.) Men were also significantly more likely to engage in the most risky behaviors than were women (again, with the exception of social risk).

The interesting question, of course, is: Why? To find out, Weber and her colleagues asked their subjects, in effect, to provide a cost-benefit analysis of each type of activity. How much risk did they perceive to be involved? And how much benefit did they think that amount of risk would bring them? When she analyzed the answers, Weber found something surprising: men weren't necessarily more risk seeking; they just valued the benefits of that risk more than the women did (the one exception, again, being the social category).

At first, said Weber, this finding may seem counterintuitive. One might expect the benefits of any given activity to be the same for men and women; a jump on a bungee cord, after all, is a jump on a bungee cord, regardless of whether it's a man at the end of the rope or a woman. But the *perceived* benefits of such an activity, she found, can be quite different, and this difference in perception can often explain why women won't take some chances that men will: they think they're not worth the risk.

### Lying and Lottery Tickets

Men and women not only perceive some aspects of the world differently; they often perceive *themselves* differently. When it comes to making mistakes, for instance, women appear to be harder on themselves than men are. For example, studies have shown that men tend to forget their mistakes more readily than women do. And mistakes appear to dog women in ways that do not bother men. In interviews, for instance, women indicate that situations involving failure affect their self-esteem more than do situations involving success; no such difference has been reported for men.

For many traits women have also been found to be less optimistic (or perhaps more realistic) than men. Various studies have shown that college men, for instance, are more likely than college women to expect to do well, and to judge their own performance favorably once they have finished their work. Some of the tasks involved in these tests were ones in which men characteristically *do* do better (such as work with geometric figures); but many were not (such as solving anagrams). Women get at least as good grades as men do—and often better. Yet, when asked what grades they *think* they will get, men are more optimistic. Even

> **Even when they lie, men and women have been shown to lie in different ways.**

when they tell lies, men and women have been shown to lie in different ways. College men tell more lies about themselves, for instance, tending to exaggerate their plans and achievements (especially when talking with women). College women, on the other hand, tend to lie to enhance another person.

This relative difference in confidence about one's prospects in the world can contribute to mistakes in tangible ways. In one telling illustration from a real-world experiment, researchers sold lottery tickets for $1.00 to men and women at a corporation. Afterward, researchers went around to the people who had bought them and asked if they'd be interested in selling the tickets back, and if so at what price. Women, they found, were willing to sell out for a much lower price than men were. On average, the women wanted just $1.33 for their tickets—not much more than the original purchase price of $1.00. The men, by comparison, wanted more than four times as much for their tickets.

> **Women were willing to sell their lottery tickets for much less than men were.**

## A Computer Error

Other types of gender-related errors are less obvious. Take, for instance, the way we use computers. Like math and war, the computer world is male dominated. After peaking in 1985 at 37 percent, the share of bachelor's degrees in computer science awarded to women has steadily fallen. Today, women receive just over 22 percent of them, or about one out of every five.

This gap intrigues the Microsoft employee Laura Beckwith, who herself recently obtained a Ph.D. in computer science. Beckwith specializes in studying the way people use computers to solve everyday problems. A few years ago she noticed that men were more likely than women to use advanced software features, especially ones that help users find and fix errors. This process of fixing errors is known as debugging, and it's a crucial step in building software programs that work.

Beckwith thought this gap could be explained not so much by a difference in ability as by a difference in confidence. When it comes to solving problems, a lack of confidence has been shown to affect not only the outcome we achieve but the approach we take. This is a subtle difference, but an important one. Among other things, self-doubters are slower to abandon faulty strategies and less likely to come up with alternatives: they stay the course.

> Self-doubters are slower to abandon faulty strategies and less likely to come up with alternatives: they stay the course.

So Beckwith, with the help of colleagues, devised a test of her own. First, she tested the confidence levels of a group of men and women by asking them whether they thought they could find and fix errors in spreadsheets filled with formulas. Then she sat them down in front of computers and had them do exactly that, working against the clock.

The key to success was using the debugging feature of the spreadsheet software. But Beckwith found that only those women who believed they could do the task successfully—that is, only those with high confidence—used the automated debugging tools. The women with lower confidence, on the other hand, relied on what they knew, which was editing the formulas one by one. This approach actually ended up introducing *more* bugs into the system than when they started.

This was puzzling. Beckwith knew from questionnaires handed out after the test that the women understood how the debugging tools were supposed to work—yet many of the women chose not to use them. Why? Once again the answer comes down to the ways men and women perceive risk. When the women in Beckwith's study did their own private cost-benefit analysis, many of them concluded that the risk of making a mistake by using the debugging tools was not worth the potential reward of fixing the bugs.

"Their perception," said Beckwith, "is that this will take me too long to learn to actually use correctly, and I'm afraid that if I started to use it, I will use it incorrectly, and it will put me further behind than if I don't use it at all."

## The Importance of Tinkering

Beckwith said the men in her study seemed more inclined to tinker with the program than the women were, and that tinkering was related to success. This squares with other research on the subject. Studies of elementary school students in math, geography, and gaming, for instance, show that boys tend to tinker and to use tools in exploratory, innovative ways. Girls are less likely to tinker, preferring instead to follow instructions step-by-step; here, too, they stay the course.

These preferences, interestingly, are reflected in the way boys and girls choose to navigate their physical world, both as children and as adults. In the United States, according to one study funded by the U.S. Department of Transportation, roughly 20 percent of the miles driven and 40 percent of the time spent driving are attributed to what researchers politely call "navigational failures"—in other words, getting lost. Moreover, the study found, when it comes to these failures, a driver's performance was largely unaffected by the driver's level of education, age, or driving experience. Old or young, experienced or not— drivers spent a lot of time being lost. But the researchers did note one factor: male drivers appeared to perform slightly better than female drivers. Not everywhere, and not at all times, but enough for the difference to be noteworthy.

### How Boys and Girls Navigate

The roots of this difference emerge early, almost from the time boys and girls let go of their parents' hands and start wandering

> Boys as young as six demonstrate that most infamous of male driving characteristics: a reluctance to ask for directions.

about the world on their own. Boys as young as six, for instance, have been shown to demonstrate that most infamous of male driving characteristics: a reluctance to ask for directions. Much of what we know about boys and girls and their ability to find their way in the world comes from Ed Cornell, a retired professor at the University of Alberta and one of the world's experts on the science of wayfinding, or how we get from point A to point B.

In one study, Cornell and his colleagues compared the navigational skills of three groups of people: six-year-olds, twelve-year-olds, and twenty-two-year-olds. Each of the participants—there were 180 in all—was led by a tester on a walk across the campus of the University of Alberta. At the end of the walk, the participants were asked to lead the way back along the same route. Cornell and his colleagues then measured the distance traveled on the return trip, noting how much of the travel occurred along the original path and how much occurred when the participants strayed off the original path.

Of all the groups, which one wandered off the path the most? By far, it was the six-year-old boys. After analyzing the results, Cornell and his colleagues noticed something else: when wandering, the girls were more likely than the boys to accept offers of help back to the correct route. "This result," concluded Cornell, "is consistent with the stereotype that, when lost, females will stop to ask for directions, whereas males muddle through various wayfinding alternatives."

There is speculation about why such differences in wayfinding exist between boys and girls. A leading explanation is that girls, from the time they are young, aren't encouraged to explore—or, if you like, to tinker—as much as boys are. For decades, researchers have studied the development of children's "home range"—that is, the distance from home that parents permit independent travel or activities by a child. A number of factors affect the size of this range. If parents are fearful, for instance, they may keep their children close by, reducing the size of the home range. City kids also may not wander as far as country kids. But in general, two trends hold true: home range for children expands rapidly between the ages of six and nine; and the range is universally bigger for boys.

## Making Maps

Research also shows that the differences between boys' and girls' mapping abilities emerge around the age of eight. This is almost exactly the time, coincidentally, that parents begin to grant boys greater freedom to roam. Up to this age, boys and girls typically have home ranges of the same size. In one study, for instance, seven-year-old girls and seven-year-old boys both had home ranges of roughly two hundred yards. But at the age of eight, a gap begins to open. By the age of nine, the boys are allowed to roam more than twice as far as girls are. Moreover, this trend holds true no matter where the kids grow up—in cities, in suburbs, or in the country.

> By the age of eight, boys have a much bigger home range than girls do.

Researchers have found that from the age of eight onward, boys begin to describe their home areas in rich and vivid terms. Their maps are more detailed, often containing twice as much information as maps drawn by girls of similar age. By the age of eleven, boys' maps become more sophisticated and show greater use of symbols and proper use of scale. Overall, boys exhibit a keener spatial awareness than girls do, even when children are asked about an environment that is equally familiar to both genders: their classroom. Boys, whether in kindergarten, second grade, or fifth grade, were more likely to be more accurate than girls in constructing models of their classroom.

The interesting thing is that these differences don't appear to be inherent; boys, in other words, aren't born better at geographic tasks than girls are. When researchers studied the subset of girls who traveled most freely and widely in their local areas, for instance, they found that these girls use about the same amount of detail in their maps as boys do.

A key part of the home-range experience seems to be the *way* kids encounter their environment. As with most kinds of learning, an active experience is much better than a passive one. It's not enough, for instance, for children to expand their home range by taking a long bus ride to school. That's because kids do what we do when faced with a long, monotonous commute: they tune out. They may stare out the window, but they don't notice their environment. And the noticing is what's important. When children are allowed to roam freely over an area and closely explore rocks, trees, creeks, and the like, they seem to develop a deeper understanding of place.

## Why Men Don't Ask for Directions

These encounters, in turn, can have a profound impact that is likely to last well into adulthood. Throughout their lives, for instance, men report having more confidence about their sense of direction than women do—even though there is little evidence that they actually *have* a better sense of direction. This difference in confidence has been documented in studies going back thirty years. In one of the earliest reports, 71 percent of men said they had a "good sense of direction." Only 47 percent of women reported the same thing. More recent studies have found similar results, with women reporting a poorer sense of direction and higher levels of anxiety about getting lost. Women tend to report more anxiety even when it comes to everyday tasks like taking shortcuts or figuring out which way to turn when coming out of a parking garage.

As Beckwith found with computing, the difference in confidence between men and women is reflected in the different approaches they take to navigating. As a general matter, men prefer a more abstract approach, navigating through the use of metrical distances, like miles, and cardinal directions, like east and west, north and south. Ask a man for directions and he might tell you, "Go one mile north, then bear east for another three miles." Researchers call this approach a "survey" strategy. Women, on the other hand, often prefer a more concrete, step-by-step approach that relies on landmarks and left-right terms. Ask a woman for directions, for instance, and she might say, "Turn right at the fire station, go down the road until you see the church, then turn left."

These are generalizations, of course. They don't apply to everybody, and they don't apply in all situations. Some women might prefer a survey strategy, and some men might use a route strategy. Depending on the circumstances, they might even switch off, using whichever strategy fits the environment. If you happen to live, as I do, in the Midwest, where open spaces abound and landmarks can be few, you might be more inclined to use cardinal directions of east and west, regardless of whether you're a man or a woman. But on the whole, researchers have repeatedly found that men and women tend to employ different strategies. And that difference helps explain why men are reluctant to ask for directions.

"There's a good reason why they don't ask for directions," says Daniel R. Montello, a professor at the University of California, Santa Barbara, who has studied the issue. People usually attribute this trait to the male ego, he says. And there's something to that, he says, at least up to a point. "But the fact of the matter is that's not the only reason."

Just like the six-year-old boys who wandered off the path in Ed Cornell's experiments, many grown men are content to wander off the path in real life. But just because they're off path, says Montello, doesn't mean they're off track.

"They don't think they're lost."

## Chapter 10

# We All Think We're Above Average

Not *long ago* a Princeton University research team asked people to estimate how susceptible they and "the average person" were to a long list of judgmental biases. Most of the people claimed to be less biased than most people. Which should come as no surprise; most of us hate to think of ourselves as average—or, God forbid, *below* average. So we walk around with the private conceit that we are above average, and in that conceit lies the seed of many mistakes.

"Overconfidence is, we think, a very general feature of human psychology," says Stefano DellaVigna, a professor of economics at the University of California, Berkeley. He has studied the ways overconfidence induces us to commit everyday errors, from signing up for gym memberships we won't use, to buying time-shares in a condominium (which we also won't use, at least as much as we think we will), to falling for teaser-rate offers on credit cards (which we will use far too much). And his research has led him to a general conclusion: "Almost everyone is overconfident—except the people who are depressed, and they tend to be realists."

> "Almost everyone is overconfident—except the people who are depressed, and they tend to be realists."

## Lessons from the Putting Green

Overconfidence traps lurk everywhere, often where you least expect them. One of my favorite examples is the golf course pro shop. If you've been in one lately, you'll know that the shops often provide a small putting green where customers can try out new putters. But these greens, it turns out, are less about the putter than they are about the one doing the putting; the greens are actually secret confidence builders that can induce people to buy more expensive clubs than they otherwise might. One experiment found that when people putted on such greens from a three-foot distance, they sank more of their putts, and as a result thought they were much better golfers than others who putted from a ten-foot distance. The three-footers, for instance, ranked themselves in the thirty-fifth percentile of skill, whereas the ten-footers ranked themselves only in the fifteenth percentile. This ego boost, as it turns out, is an effective sales tool; those who putted on the shorter green thought they should buy higher-end equipment than those who putted from a longer distance.

Overconfidence can lead to much bigger mistakes than paying too much for a set of clubs. But before we talk about these, let's start with a question: If, over the last three years, you had to pick one stock to invest in, which one would it be? My nominee would be NutriSystem, the diet company based in Horsham, Pennsylvania. It sends low-calorie, prepackaged meals through the mail to millions of people who want to lose weight. And the company (if not its customers) has been enormously successful. Through 2006, the stock's compound annual average return over three years was a stunning 233 percent. That made it one of the top performers not only in the diet business but on all of Wall Street. An investor who put $1,000 into NutriSystem shares at the end of 2003 would have had a whopping $36,855 just three years later.

## Getting Rich off Fat People

The interesting thing about NutriSystem, aside from its stock performance, is its customers: they are overconfident. The typical Nutri-System customer is what the company, in a presentation to Wall Street analysts, calls a "serial dieter"—that is, somebody who has tried but failed to lose weight. Most of them (80 percent) are women. On average, they are forty-four years old and weigh 210 pounds. Most start out wanting to lose 60 pounds, but end up losing only about 20. Typically, after ten or eleven weeks, they give up and drop out of the program.

Why, then, if its customers fail has the company succeeded? The answer is that NutriSystem, like a great many corporations, has learned to capitalize on our overconfidence. It banks not on what people *will* do but on what people *believe* they will do. In Nutri-System's case, this belief is abetted by the company's advertisements (just as the putters' overconfidence is abetted by short greens). Typically, its ads feature celebrity athletes like the former National Football League quarterback Dan Marino, who says he lost twenty-two pounds, and Mike Golic, the former defensive lineman turned ESPN analyst, who says he lost fifty-one pounds.

Look closely at the ads, though, and you will notice in small print a disclaimer consisting of three important words: "Results not typical." You'd think this would be a tip-off to potential customers that losing a lot of weight through the company's program is unlikely. But it isn't. To prospective dieters, it doesn't matter if the advertised results aren't typical, because most people think *they're* not typical. They're above average—and their results will be, too.

## Why We Pay Not to Go to the Gym

In this sense, NutriSystem isn't in the diet business; it's in the hope business—and so, if you think about it, are banks and gyms and many other businesses, too. Health clubs are a multibillion-dollar industry in the United States. By one count, nearly thirty-three million people pay some $12 billion a year to work out. Typically, most of them enroll through membership plans. But are they wasting their money?

To find out, DellaVigna and a Berkeley colleague, Ulrike Malmendier, looked at records from three U.S. health clubs that detailed the day-to-day attendance of nearly eight thousand members over three years. Like you or me, these health club patrons typically had three membership choices:

1. An annual contract
2. A monthly contract
3. A pay-per-visit option, often in the form of a ten-visit pass

Which one would you choose?

If you're like most people, you'd probably choose the annual or monthly contract—and you would probably overpay. That's because gym members, like dieters, tend to be overconfident. They believe they will go to the gym much more often than they do. In fact, DellaVigna and Malmendier found, gym members go to the gym only about half as often as they expect to—four to five times a month instead of the ten times per month they expect they will.

> Gym members go to the gym only about half as often as they expect to go.

As a result, gym members with contracts overpay for the visits they do make. On average, DellaVigna and Malmendier found, members overpaid by about $700 apiece. Overpaying like this isn't universal—but it's close. Some 80 percent of monthly members would have been better off with the per-visit plan. So why do they overpay?

"Basically, you overestimate your self-control," says Della-Vigna. "You confuse what you *should* do with what you *will* do." In other words, you behave a lot like a NutriSystem customer.

Interestingly, contracts like the ones DellaVigna studied haven't always been the norm in U.S. health clubs. In the 1950s, many health clubs operated under a pay-per-use system. Typically, they doled out coupons, which customers redeemed at the entrance. But as clubs converted to electronic card systems that enabled them to monitor their members' attendance patterns, the coupons were replaced by contracts.

DellaVigna believes that over time health clubs, like other businesses, have wised up; they have learned to design their contracts to take advantage of their customers' built-in optimism. "In many ways the firms are taking advantage of that knowledge," he said, "and consumers should be smarter in realizing that."

But this is easier said than done. As in the case of the short putting green, it's not always easy to recognize when our overconfidence is about to get us snookered. We are often lulled into complacency by initial costs that seem low. There is a reason, after all, why hotel rooms in Las Vegas are cheap and why cell phone calling plans give you "free" minutes: both the casinos and the cell phone companies know you will overestimate your self-control. You will gamble and talk more than you think you will, and when you do, they will profit and you will not.

### How Credit Cards Play on Your Overconfidence

This is also the reason why credit card companies offer teaser interest rates—low initial rates that jump after an introductory period of, say, six months. According to the Federal Reserve, the average credit card debt per household in the United States is about $8,500. Maybe you have some of this debt yourself and would like to cut your interest payments. Say you walk down to your mailbox today and find three credit card offers waiting for you:

- The first one offers an interest rate of 6.9 percent for the first six months and 16 percent thereafter.

- The second one offers 4.9 percent for six months and 16 percent thereafter.
- And the third one offers 6.9 percent for six months and 14 percent thereafter.

Which one would you choose?

One study tracked nearly two million direct-mail solicitations issued in the 1990s by a major credit card issuer in the United States. In all, these solicitations yielded just over fifteen thousand new credit card accounts. Which means 99 percent of the people who received them did what you probably do with credit card offers: they threw them in the trash. So why do banks bother?

The answer lies, in part, in understanding the behavior of the remaining 1 percent. For this group, the hands-down winner was choice No. 2, which offered the lowest teaser rate for the first six months: 4.9 percent. It was preferred by a margin of more than two to one. Why?

The answer, once again, appears to involve overconfidence. Just like the fat people at NutriSystem who hoped to lose sixty pounds but ended up losing only twenty, credit card applicants hoped to pay off more of their loan balances before the higher interest rate kicked in. As a result, those who responded to the credit card offers placed more importance on the low initial interest rate of 4.9 percent—and not on the subsequent high interest rate of 16 percent. But, as with NutriSystem, the optimism doesn't pay off. Most people didn't pay off their debt in the introductory period. Instead, they carried a balance into the higher-interest-rate period.

### Knowing Your Limitations

If you've ever seen the movie *Magnum Force*, you no doubt remember Clint Eastwood's role as the tough-guy detective Harry Callahan. At the end of the movie Eastwood has the bad guy right where he wants him—when he utters one of his more famous lines: "A man's got to know his limitations."

Good advice, that.

When it comes to judging just how overconfident someone is, social scientists have devised a term for knowing one's limitations: it's called "calibration." Calibration measures the difference between actual and perceived abilities. If you're as good as you think you are, then you are said to be well calibrated. If you are *not* as good as you think you are, then you are said to be poorly calibrated.

Most of us tend to be poorly calibrated, even (and perhaps especially) when it comes to important skills, like those we need to do our jobs. The U.S. Army discovered this years ago when it asked soldiers at Fort Benning, Georgia, the military version of the layman's question "How good a shot are you?" Most of the soldiers, naturally enough, thought they were pretty good shots. They predicted they would score well on the Army's annual qualification with M-16 rifles. Then the Army asked the same soldiers to step onto a firing range. After the shooting was done, the soldiers' scores were tallied and compared with their predictions. The results were not good: 75 percent of the soldiers predicted they would hit more targets than they did. In addition, more than one out of every four soldiers shot so poorly that they failed to qualify. The soldiers "generally overestimated their actual performance," concluded a report of the results, "and were biased heavily toward predicting success."

> Most of us tend to be poorly calibrated— we're not as good as we think we are.

Oddly enough, the Army noted, the predictions by one group of soldiers proved to be on target. That group? The poorest shots. To be sure, this was a small group. Of the 153 soldiers participating in the annual qualification, only 5 predicted that they would fail to qualify. But those 5 were dead-on. Three of them did, in fact, fail to qualify; the other 2 just barely passed.

"Those who predicted failure," the researchers noted drily, "were quite accurate."

Researchers have found equally poor calibration for people engaged in other tasks, regardless of factors like income, intelligence, and education. Not long after the Army tested the soldiers at Fort Benning, for instance, researchers conducted a similar test of students at the University of Wisconsin, Madison. Instead of having the students shoot at targets, though, the researchers asked them to do the academic equivalent: read one paragraph of text and then rate themselves in terms of their confidence in their ability to draw the correct inferences from it. Then the students were tested on what they had read. As you might expect, the students, like the soldiers, didn't do so hot.

"Calibration was strikingly poor," said the researchers. "Our readers were unable to distinguish between what was understood and what was not understood." Go, Badgers.

## What Weather Forecasters Can Teach Us

There is one group, though, whose members are remarkably well calibrated: weather forecasters. The reason for this can be traced to—of all places—Roswell, New Mexico, the UFO capital of the United States, and to a young weather forecaster who worked there many years ago named Cleve Hallenbeck. After school Hallenbeck bounced around from job to job—railroad man, grocery clerk, schoolteacher. Eventually, he landed a job with the U.S. Weather Bureau, as it was then known, which sent him to one of the more remote spots a weatherman could hope to find: the Pecos valley of New Mexico.

Then, as now, the Pecos valley was alfalfa country. Most crops were irrigated from wells, and farmers needed to know when they would have to use their own water, which cost them money, or when they could wait for rain, which cost them nothing. They also needed to know when it would *not* rain, because the farmers needed long dry periods to cut and cure the alfalfa hay. But simply telling the farmers there was a "chance" of rain wasn't good enough; they had too much at risk. They needed to know if that chance was a 50 percent chance, a 75 percent chance, or a 100 percent chance. So Hallenbeck began including these probabilities with his forecasts for the Pecos valley. That was in 1920.

It took a while, but the use of probability statements caught on. Hartford, Connecticut, began using them in its forecasts to the general public on a regular basis in 1954. San Francisco did likewise in 1956, and Los Angeles followed suit in 1957. In 1965, the entity now known as the National Weather Service initiated a nationwide program in which precipitation probabilities were regularly included in all public weather forecasts. The program represented the first time that probabilities in any field of application were issued on such a large scale, and the program continues, largely unchanged, to this day.

As a result, weather forecasters have established a long track record of predictions made in quantified terms. They also, of course, have a record of the actual results: they can tell whether it rained on a certain day or not. When the predicted results and the actual results are laid side by side, weather forecasters come out remarkably well. One analysis of more than 150,000 forecasts made over a two-year period found that weather forecasters were nearly perfectly calibrated. When U.S. weather forecasters predicted a 30 percent chance of rain, for instance—as they did more than 15,000 times in this study—it rained almost exactly 30 percent of the time.

One analysis found that weather forecasters were nearly perfectly calibrated; when they predicted a 30 percent chance of rain, it rained almost exactly 30 percent of the time.

### The Power of Feedback

Why are weather forecasters so well calibrated when the rest of us aren't? The answer involves one of the best cures for overconfidence: quick, corrective feedback. What's feedback? Basically, it's a signal. Feedback sends back to the user information about what action has actually been done and what result has been accomplished. Feedback is a well-known concept in the science of information and control theory; it's why phones have dial tones and make those little beeps when you push the buttons—the feedback lets you know whether you've done something correctly.

Feedback is a powerful way to shape human behavior. It's why slot machines pay out immediately. Casinos want you to keep doing what you're doing, which is gambling. So when you win, you don't get a check in the mail a month later: you get your money *now*. Instant feedback.

In situations where overconfidence is high, feedback is often low, either in quantity or in quality—or both. Remember our example about people paying not to go to the gym? After he looked at how much money most gym members were wasting, one question

> In situations where overconfidence is high, feedback is often low.

kept lingering in DellaVigna's head: "Why would adults make this mistake?" After all, he reasoned, they've had a lifetime to learn about their self-control. The answer involved weak feedback.

Like any signal, feedback can be strong or weak. If you dial a phone number and get a busy signal, that's strong feedback: the phone is busy, and you know instantly that it's busy—there's no guesswork involved. But when you don't go to the gym, the feedback is weak: you're not really sure *why* you didn't go. And when feedback is weak, it is easy to ignore or distort the signal that is sent. In the case of not going to the gym, the signal is that we're lazy. But we don't like that signal, so we ignore it.

"We try to reject that feedback because that really hurts our feelings," says DellaVigna. So we distort the feedback by attributing our mistake to some cause other than laziness. "Every time that we do something like didn't go to the gym, we can always attribute it to something else—my kid was sick, or all of that."

Certain occupations are ideally suited to feedback distortion. Take, broadly, the field of corporate finance. When someone decides to buy a stock or even an entire company, it may not be clear for years that the move was a mistake. By then, those responsible for it have often moved on to other jobs at other companies. And those who are still around can do the equivalent of blaming the sick kid: they can attribute the error to any number of causes.

### Warren Buffett's Biggest Mistake

An interesting case involves the Dexter Shoe Company of Dexter, Maine. The billionaire investor Warren Buffett, along with his longtime partner, Charlie Munger, bought the company for $433 million in 1993. At the time, Buffett thought it was a great deal. "It is one of the best-managed companies Charlie and I have seen in our business lifetimes," he told the shareholders of his firm, Berkshire Hathaway.

By 2008, though, Buffett had changed his tune.

"To date," he told shareholders, "Dexter is the worst deal that I've made."

What happened? Within a few years of the deal, Dexter's competitive advantage, which Buffett had thought durable, had vanished.

"But that's just the beginning," he said. Buffett paid for the company not in cash but in Berkshire stock—25,203 shares of

Berkshire class A stock, to be exact. And by using the stock, he said, "I compounded this error hugely." The move made the cost to Berkshire shareholders not $433 million but $3.5 billion. "In essence, I gave away 1.6% of a wonderful business—one now valued at $220 billion—to buy a worthless business."

This mistake is interesting for a couple of reasons. The first is that Buffett would acknowledge it (not many CEOs would). The second is that he has a specific explanation for the error (miscalculating the durability of the competitive advantage). This indicates that he has spent some time thinking, in essence, about why he didn't go to the gym—and was honest with himself about the feedback he got. The third, and perhaps least obvious point, is that Berkshire has virtually no turnover—either at the top or down below. Buffett has been at the helm from the company's inception, and the chief executives below him who run Berkshire's forty-some business units almost never leave. ("How many CEOs have voluntarily left us for other jobs in our 42-year history?" Buffett asked in his 2006 letter to shareholders. "Precisely none.")

> Berkshire Hathaway's executives stick around long enough for feedback to reach them, allowing them to learn from their mistakes. Between 1964 and 2007, the company reported an overall gain of 400,863 percent.

In short, the company's executives stick around long enough for the feedback to reach them, allowing them to learn from their mistakes. This may help explain why, over the long haul, Berkshire's performance has been so spectacular. Between 1964 and 2007, the company reported an overall gain of 400,863 percent. By comparison, the Standard & Poor's 500 Index with dividends included, a broad measure of the stock market's performance, gained only 6,840 percent.

Despite such an impressive performance, Buffett exhibits

none of the overconfidence one might expect from one of the world's richest men. As he noted in his 2007 letter, "I'll make more mistakes in the future—you can bet on that."*

## The Illusion of Control

Oddly, as tasks get harder, the degree of overconfidence tends to go up, not down. You'd think it'd be the other way around, but it's not. Overconfidence is typically most extreme with tasks of great difficulty. Even when people are given a job that's nearly impossible to do—like trying to tell the difference between drawings by Asian children and those by European children—they think they will perform much better than they do.

So strong is our belief in our own abilities that we often believe we can control even chance events, such as flipping a coin or cutting a deck of cards. In a famous series of experiments conducted years ago, Ellen Langer, now a professor at Harvard University, demonstrated this tendency on a group that ought to know better—students at Yale. Langer had Yale undergraduates play a card game against a colleague. The game was simple: each person drew a card from the deck; whoever drew the higher card won that round. On each round students could place a bet, ranging from nothing to twenty-five cents.

But the game was rigged. Some of the Yale undergraduates cut cards against a colleague who was a sharp dresser and exuded competence; others played against a schlub in an ill-fitting sport coat. In either case, the chances of drawing the high card were the same; the deck of cards, after all, doesn't care who the players are. But the

---

*It is worth noting that Buffett is a passionate bridge player and that bridge players, as a group, tend to be very well calibrated. "You know," Buffett once cracked, "if I'm playing bridge and a naked woman walks by, I don't even notice." See Blackstone (2008) and Keren (1987).

students cared, and that's the important point. When the students cut cards against the schlub, they felt much more confident that they would pick the higher card. Their bets, in turn, reflected this confidence: when playing against the schlub, the students consistently bet more money than they did when playing against the sharp dresser.

Langer found a similar effect when she asked students to predict the outcome of a coin toss. This game, too, was rigged. A colleague would flip the coin, and the student would call out "heads" or "tails" while the coin was in the air. What the students didn't know is that the flipper called out the results of the coin toss in a predetermined order. Some students were told that their first few guesses were right; others were not.

This initial success had a powerful influence on the students' confidence. After a while, those students who believed they had guessed correctly in the beginning became convinced that they were skilled enough at guessing heads or tails that they could actually beat the odds and make successful predictions more than half the time. Even more interesting, though, were the comments made by the students afterward; 40 percent of them felt they could actually improve their performance with practice. Langer dubbed this phenomenon "the illusion of control."

## Information Overload

What might explain the persistence of such an illusion? Part of the answer lies in the beguiling power of information. The more we read (or see or hear, for that matter), the more we think we know. But, as has long been observed, that isn't necessarily so. Often what happens is that we don't grow more informed; we just grow more confident.

Summaries of information, for instance, often work as well as—and sometimes even better than—longer versions of the same material. In a series of experiments, researchers at Carnegie Mellon University compared five-thousand-word chapters from college textbooks with one-thousand-word summaries of those chapters. The textbooks varied in subject: Russian history, African geography, macroeconomics. But the subject made no difference: in all cases, the summaries worked better. When students were given the same amount of time with each—twenty to thirty minutes—they learned more from the summaries than

> **Students learned more from summaries than entire chapters.**

they did from the chapters. This was true whether the students were tested twenty minutes after they read the material or one year later. In either case, those who read the summaries recalled more than those who read the chapters. (So if you relied on CliffsNotes in college, take heart.)

But deep down we don't want to believe this. We seem to have an innate desire to overload ourselves with information—whether it helps us or not. Indeed, "information overload" has become a cliché. Information is constantly piped to us—from the video screen in the back of the cab to the TV in front of the StairMaster to the keyword alerts that pop up on our e-mail. We all crave information, though nowhere, perhaps, is it prized so much as at the racetrack.

## Horse Sense

"Horse players are information junkies," says Jill Byrne, who should know. She is one of two handicappers at Churchill Downs, home of the world-famous Kentucky Derby. She also appears as a handicapper on the horse-racing television network TVG, which is seen by millions of households across the country. It is her job to tell gamblers which horse is likely to win. Like stock analysts and others who make a living trying to predict the future, she is often wrong. Byrne's track record at Churchill shows a win rate of 32 percent, which, though short of 50 percent, is still very good by handicappers' standards.

She attributes her skill to her upbringing. When she was a child, everything she did revolved around horses. At the age of two, she began riding horses on the family's horse farm in Charlottesville, Virginia. At seven, she was running the barns along with her sister. When she was twelve, her father, a horse trainer based in New York, put her on her first racehorse. During Jill's senior year of college her parents divorced. Jill dropped out of school and moved to New York to be with her father, who spent much of his time at Belmont and Saratoga. "At the time," she said, "it seemed like the glamorous thing to do."

In a typical year, she will handicap thousands of horses, in races large and small. For a major race like the Kentucky Derby, which is run on the first Saturday in May, Byrne begins her research in October. She will ultimately handicap three or four picks. For a given horse, she can easily analyze hundreds of discrete pieces of information. For instance, she will analyze the past performance not only of the horse in question but of that horse's parents, its grandparents, and even its brothers and sisters. Did they do better in long races or short? What type of surfaces do they prefer—turf or dirt? What if it rains—how well do they do in the mud? She'll also look at the horse's competition to see whether it is up against a tougher class. "How much speed is in that race? Does the horse like to run in the front? Are there a lot of other horses in the race that like to do that, too, that may compromise his chances? Or is he what I would call the lone speed? Does he get out there by himself, and have a better chance then of not being pressured and can finish better?"

Answering these questions takes time. Each pick, she estimates, requires maybe fifty hours of study. But how much good does all of this information do?

Years ago, Paul Slovic, then at the Oregon Research Institute, posed a similar question. He tested the predictive abilities of professional horse-racing handicappers like Byrne. In Slovic's experiment, the handicappers were allowed to form their judgments by using "past performance charts" like those available in the horse-racing bible, the *Daily Racing Form*. Such charts gave nearly a hundred pieces of information on each horse and its history.

In Slovic's study, eight handicappers made predictions for forty races. (These were real races, by the way, allowing Slovic to compare the predictions with actual results.) At first, the handicappers were allowed to use only five pieces of information per horse from the charts when making their predictions. They could use any five they wanted—the weight of the jockey, for instance, or the percentage of races in which the horse finished first, second, or third. Then the handicappers were asked to make the same kinds of predictions using ten pieces of information per horse. Then twenty pieces per horse. Then forty.

Did using more information make the handicappers' predictions more accurate?

No.

Their accuracy was no better with forty pieces of information than it was with five. But—and this is an important but—using more information *did* increase the handicappers' confidence. This increased substantially, from less than 20 percent with five pieces of information to more than 30 percent with forty pieces of information.

> Handicappers' accuracy was no better with forty pieces of information than it was with five. But—and this is an important but—using more information did increase their confidence.

## Executive Decisions

This tendency, of course, is not unique to those who make their living at the track. Corporate executives routinely display overconfidence in their judgments about the thing they think they know best: their businesses. Listen, for instance, to Donald Tomnitz, the chief executive officer of the nation's largest home builder, D. R.

Horton, based in Fort Worth, Texas. In December 2005, Tomnitz, a former banker and Army captain, boldly said of his company, "We can earn our way through any economic cycle, except one like the Great Depression."

Many people believed him, too. Investors poured billions of dollars into Horton, pumping its stock price to near-record levels. Another Great Depression never arrived, but a housing bust did. Soon, Horton's earnings declined more severely than almost anyone imagined. By the summer of 2007—just nineteen months after Tomnitz's boast—Horton reported the first quarterly loss in its fifteen-year history as a publicly traded company. And it wasn't a little loss, either. It was a whopper: nearly $824 million. More losses would follow, sapping Horton's value. At the time of Tomnitz's boast, Horton's shares traded for about $36 apiece. By the summer of 2008, they had fallen by two-thirds to about $12.

For years, Paul Schoemaker, a professor at the University of Pennsylvania's Wharton School, and J. Edward Russo, of Cornell, have studied the overconfidence of executives like Tomnitz. Overconfidence in corporations is so common, they say, that it constitutes "a hidden flaw in managerial decision making."

How can they tell? Over the years they have given a "confidence quiz" designed to measure what they call metaknowledge: an appreciation of what we do know and what we do not know. In essence, it's a calibration test. The quiz typically has ten questions, and those who take it are asked to answer each question with an estimate that has a certain range of confidence, say 90 percent. For instance, one question might be: How long is the Nile River? If you're 90 percent sure the answer is between 500 and 600 miles, you would write down "500–600 miles." I've included a similar sample quiz at the end of this chapter. If you want, give it a try.

Schoemaker and Russo have given these quizzes to workers in a variety of industries. The nature of the quizzes varies. Sometimes the questions test general knowledge by asking how many patents were issued in a given year, or how far it is from New York to Istanbul. Often, though, the questions are tailored to an industry and sometimes even to a specific firm. You'd think this would give the test takers an upper hand; after all, managers are expected to know more about their company or industry than about the world at large. But in the end, the tailoring doesn't make much difference.

"Every group believed it knew more than it did about its industry or company," they concluded.

In a test of managers in the advertising business, for example, the confidence rating was placed at 90 percent. This meant the test takers should have been wrong 10 percent of the time; instead, they were wrong 61 percent of the time. In the computer industry, test takers should have been wrong 5 percent of the time; they were wrong 80 percent of the time. (In both cases, by the way, the tests were tailored to the respective industries.)

Russo and Schoemaker have administered such quizzes to more than two thousand people. Of these, Schoemaker said, more than 99 percent proved overconfident.

• • • • • • • •

## Confidence Quiz

• • • • • • • •

For each question below provide an answer within a range that you are 90 percent sure is correct. For instance, if you are 90 percent sure that the correct answer to a question is somewhere between 1 million and 2 million, write down "1 million–2 million." The answers are at the bottom of the page.

1. Average number of miles driven by an American in 2005?
2. Number of marriages in United States, 2003?
3. Average number of times adults report having sex each year, worldwide?
4. Number of U.S. states where more than 14 percent of adults are obese?
5. Number of bachelor's degrees earned in the United States, 2003?
6. Median net worth of a U.S. family, 2001?
7. Total land area of the United States in 2000, in square miles?
8. Number of state and federal prisoners in United States, 2003?
9. Number of births in United States, 2003?
10. Number of deaths in United States, 2003?

• • • • • • • •

Answers: 1. 13,657; 2. 2,187,000; 3. 103; 4. 50; 5. 1,348,503; 6. $86,100; 7. 3,537,438; 8. 1,409,280; 9. 4,091,000; 10. 2,444,000

## Chapter 11

## We'd Rather Wing It

If *our ability* to know how good—or bad—we are at something improved over time, that would be one thing. But this skill, sadly, doesn't necessarily improve with experience. One of the best illustrations of this comes from the Professional Golfers' Association. In the late 1980s, the field staff of the PGA quietly conducted a test of its golfers' putting abilities. Putting, after all, is a key part of the game, accounting for 43 percent of all strokes. Among other things, the PGA wanted to know what percentage of six-foot putts is made by the best golfers in the world. With funding from *Sports Illustrated*, the PGA measured putts at each of fifteen tournaments in the latter half of 1988. At each tournament the field staff chose one green with a smooth and relatively flat surface. Then, for the four days of competition, every putt was measured.

### Lessons from the Putting Green—Part II

In all, the PGA cataloged records on 11,060 putts. (For statistical reasons, 2,593 tap-in putts of eighteen inches or less were eliminated from consideration.) Of the remaining putts, 272 were six-footers. So what percent of these did the best golfers in the world sink?

The answer, it turns out, was just over half: 54.8 percent, to be exact.* The number itself was not so surprising. The United States Golf Association, which is the governing body for golf in the United States, had measured putting from specific distances at the 1963, 1964, and 1988 U.S. Opens. The results of those studies mirrored the PGA's findings.

The interesting part was the reaction from the professionals on the PGA tour. Most of them guessed that at least 70 percent of their six-foot putts would drop. Tour rookie Billy Mayfair, a former U.S. Amateur champion and a very good putter, thought the success average was above 80 percent and put his own average "around 91 or 92%." Veteran Dave Barr, who was more typical, said, "If you aren't making at least 85% of your six-footers, you aren't making any money." Told that the actual average was 54.8 percent, Barr said, "I don't believe it."

Most people don't, either; as we've seen, we all think we're a little bit better than we are. But if you look closely at the track records of many so-called professionals, it turns out that *they're* not all they're cracked up to be, either. When it comes to certain tasks—notably those involving judgment or prediction—their performance is often worse than they would have you believe. In one study, information from a test used to diagnose brain damage was given to a group of clinical psychologists and their secretaries. The psychologists' diagnoses were no better than their secretaries'.

> The track records of many professionals are not what they're cracked up to be.

*The PGA study yielded other interesting results. Curiously, when a golfer was putting from the same distances, the success rate for putts for pars was higher than the rate for putts for birdies—in some cases, much higher. At a distance of five feet, for instance, it was 25 percent higher. This suggests that psychological pressure affects even the best of golfers. A study of free throws made by professional basketball players reached a similar conclusion. See Camerer (1998).

Even worse are those people many of us rely on for financial advice: securities analysts. When researchers looked at their ability to predict the earnings of the companies the analysts followed, their record was not only dismal—it got worse over time. In 1980, analysts were wrong 30 percent of the time. In 1985, they were wrong 52 percent of the time. And in 1990, they were wrong a whopping 65 percent of the time.*

Similarly dismaying results come from studies that compared professionals' predictions with those made by actuarial models (essentially, computers). There have been about a hundred such studies, according to Colin Camerer, a professor at Caltech who has reviewed them. "Experts did better in only a handful of them," he concluded. The studies he reviewed covered a variety of fields—college admissions, recidivism of criminals, medical diagnosis. In some cases the "experts" were more accurate than novices—but rarely were they better than simple statistical models.

"The depressing conclusion from these studies," he wrote, "is that expert judgments in most clinical and medical domains are no more accurate than those of lightly trained novices."

Findings like this ought to generate humility. But they don't. One test of political experts' abilities to predict world events found that "experts and non-experts alike were only slightly more accurate than one would expect from chance." The key difference between the two was in their respective levels of modesty.

> Even in the face of evidence to the contrary, experts tried to "convince themselves that they were basically right."

---

*Analysts' opinions are usually biased in one direction: about 95 percent of the time, they recommend that investors buy or hold stocks; almost never do they utter the *S* word: "sell." To temper the Pollyannas in its ranks, Merrill Lynch, the nation's largest brokerage firm, in 2008 began requiring its analysts to assign an "underperform" or "sell" rating to 20 percent of the stocks they cover. See Anderson and Bajaj (2008).

"Most experts thought they knew more than they did," concluded the study. Even in the face of evidence to the contrary, experts tried to "convince themselves that they were basically right."

## Practice, Practice

Given the checkered performance of many professionals, it seems reasonable to ask the question: What *really* makes an expert expert? When the U.S. military asked this question, it found that many of its top guns were actually deep thinkers. Like chess masters and other super-experts, the pilots possessed the ability to swiftly grasp the impact of an event five or six moves into the future—that is, they could think deeply into the problem, and do it quickly. But how did they develop that ability?

Largely by developing vast memories, says K. Anders Ericsson, a professor of psychology at Florida State University. Ericsson is an expert on experts. For more than thirty years he has studied expertise in a variety of fields—from waiters and chess players to pilots and musicians. He has found that experts, no matter their field, usually have certain things in common. For one, they start young. Most world-class performers were seriously involved in their fields before the age of six. For another, innate abilities—either physical or mental—don't matter as much as people think they do. IQ tests, for instance, have been remarkably unsuccessful in accounting for individual differences in performances in the arts and sciences and advanced professions. And, with the exception of height, there is little firm evidence that innate characteristics are required for healthy adults to attain elite performance in sports.

> Experts practice—a lot. No matter the field, it takes about ten years of sustained effort to become a world-class expert.

But what does matter is practice. Experts practice—a lot. No matter the field, it is generally agreed that it takes about ten years of sustained effort to become a world-class expert. One group of experts studied by Ericsson and his colleagues is violinists. By the age of twenty, the best group of young and middle-aged violinists had spent over ten thousand hours in practice. By comparison, two groups of less accomplished violinists of the same age had logged twenty-five hundred and five thousand hours, respectively.

## A Big Library of the Mind

But not just any practice will do. Experience and expertise are not the same thing; simply repeating the same task over and over, says Ericsson, is no guarantee you will get better. Instead, the practice needs to be directed toward improving the memory of the performance. When performed correctly, prolonged, deliberate practice produces a large body of specialized knowledge—a library, if you will —in the mind of the person doing the practice. This is important because having a big library allows an expert to quickly recognize patterns that others don't. This ability was identified years ago in classic studies of chess players. In the studies, two groups of chess players were shown glimpses of a chessboard in mid-game. One group was composed of the super-elite: chess grand masters. These are the best players in the world, and they have typically logged some thirty thousand hours of playing time. The other group was less experienced, but still no slouches: they were chess experts, with about three thousand hours of playing time. When the chess masters were given a glimpse of the board, they were able to recall the positions of every piece nearly perfectly. But the recall of the less experienced players was less accurate; they got the positions right only between 50 percent and 70 percent of the time.

What accounts for the difference? It's not that the grand masters simply had better overall memories. Researchers know this be-

> Pattern recognition is the hallmark of expertise, allowing experts to anticipate events and respond quickly.

cause both groups of players were subsequently retested. This time, the pieces on the board were scrambled into positions that made no sense. When this happened, the grand masters had no better recall than the experts. In other words, the grand masters had superior memory for the positions of the pieces *only* when the positions made sense—that is, when they were part of a pattern recognized in the big library.

In many cases, the knowledge of patterns acquired by experts is so deep, their libraries so vast, they are able to create in their heads mental models of how a sequence of events should play out and quickly—almost instantly—detect problems. Chess players at the master level, for instance, can play while blindfolded with only a minor reduction in their chess ability. And expert pianists are able to fix notational errors in music on the fly, automatically correcting the music back to what the genre would have predicted.

## Cognitive Maps

More than seventy years ago a series of experiments was conducted by the late Edward Tolman, a professor at the University of California, Berkeley. Tolman was a giant in his field, a founding father in the experimental study of animal cognition. He was curious about what goes on inside the animal; it was not enough for him to know that under a certain circumstance, an animal might choose this response or that. He wanted to know why. Since we can't ask animals why they do certain things, Tolman devised a series of tests that would show him.

In one, he turned rats loose in a special maze. The maze offered a single, though roundabout, path to some food. A more direct route to the food would have saved the rat time, but no such shortcut was available. The rats were given five tries in this maze. Then Tolman repeated the experiment. But this time he modified the maze. The food was kept in the same spot as before, but the original route to the reward was blocked. Instead, a series of paths radiated out from a central area, like spokes from a hub. One of these spokes contained the shortcut to the food. The question was, would the rats choose the shortcut?

The answer, by and large, was yes. More than a third of them chose it from among the eighteen alternatives—far more than chose any other spoke. The rats, it seems, were able to view the maze much as master chess players view a chessboard; they were able to form a mental representation of the situation they faced and determine the best course of action from many possible options. They were able to form what Tolman called a "cognitive map" of the situation, complete with the objects, pathways, and rewards that were in it.

People do the same thing. Indeed, much of life, arguably, boils down to finding the shortcut to the cheese. But how, exactly, do we do this? Most of us aren't experts. We don't have a vast library to tap. We haven't practiced for thousands and thousands of hours. And we aren't even particularly deep thinkers. So how do we go about solving the 1,001 problems we face each day? We create our own cognitive maps—though not in the disciplined, structured way an expert might. Ours are a bit more haphazard—less like a Rand McNally version and more like one drawn on a cocktail napkin after a few drinks.

An insight into this process comes from Steve McConnell, a software consultant in Bellevue, Washington. When Steve was in the seventh grade, his art teacher offered a deal to the class: anyone who followed his instructions would get a grade of at least a B, regardless of artistic talent. The teacher, a 220-pound former marine, drove this point home at least once a week, reminding his students of the proposal. Nonetheless, Steve was amazed at how many of his classmates didn't follow the teacher's instructions—and didn't get at least a B. Judging by the quality of their work, he said, it didn't appear that their failure to follow instructions was due to conflicting artistic visions. Instead, recalled McConnell, "they just felt like doing something different."

A lot of us feel like doing "something different." Studies have shown that people do not like to read instructions, and much of what we do read we either ignore or don't understand. In one test, for example, twenty-four adults were asked to wire a common household electrical plug. Only ten of the twenty-four even bothered to look at the instructions. And of those ten, seven consulted the instructions only to check the color coding for the electrical wires; the rest of the information was ignored. Not surprisingly, most people flunked this test. Of the twenty-four adults, only five wired the plugs successfully.

> Twenty-four adults were asked to wire a common household electrical plug; only five did so correctly. Most didn't bother to read the directions.

Even when the instructions are unusually important, people tend to pay them little attention. One study, for example, assessed the ability of mock jurors to remember the contents of the judge's instructions. Their recall was terrible: they remembered only 12 percent of what the judge had told them.

Instead, people prefer to bushwhack. As the authors of the electrical plug study noted, "Even in the case of quite unfamiliar tasks, people seem to prefer to act rather than reflect."

### Hitting the Nail on the Head— and in the Heart and the Neck

One of my favorite illustrations of this principle involves nail-gun injuries. Nail guns typically use blasts of compressed air to drive nails into wood—or, as it often turns out, into human flesh. According to the U.S. Centers for Disease Control and Prevention, there are about thirty-seven thousand nail-gun injuries a year. And what a spectacular collection of injuries these are. Typically in such accidents, a hand or a finger is impaled by the nail. But many times, the nail misses its mark by an even wider margin. One man, for instance, was nailed in the carotid artery. Another, a teenager, was shot in the heart. And one fifty-year-old nail gunner even shot himself in the head—twice! He showed up in the emergency room complaining of a severe headache. Sure enough, after doctors X-rayed his head, they spotted two nails in his brain. (All three of these victims, we should note, were treated successfully.)

More interesting, the number of these injuries has soared. Between 2001 and 2005, according to the CDC, the number of these injuries had nearly doubled. Why? Certainly sales had something to do with it. As nail guns became cheaper and more available, injuries were bound to increase.

> The federal government found that professional nail gunners were not getting hurt—It was the do-it-yourselfers.

But when the CDC looked into the issue, it found another explanation: It wasn't the professional construction workers who were increasingly being hurt; their injury numbers were stable. No, it was the do-it-yourselfers who were being maimed by the thousands.

### The Seven-Hundred-Page Owner's Manual

The preference do-it-yourselfers have for bushwhacking is understandable. The industrial boom that followed World War II brought us not only more things but more complex things. Irving Biederman, a psychologist who studies visual perception, estimates that there are thirty thousand "readily discriminable" objects for the average adult. Donald Norman, a cognitive psychologist, pegs the number at closer to twenty thousand. Either way, it's a lot. And most of them come with instructions. Even our *clothes* now come with instructions. Before the Cissell Manufacturing Company introduced its first tumble dryer in 1951, for instance, no one knew whether or not it was okay to tumble dry a sweater; it went on the clothesline with the rest of the family's wash. But in 1971 the federal government began requiring tags on clothes to explain how they should be cleaned and dried, and they've been itching the backs of our necks ever since.

Some instruction manuals now run the length of a novel—or two. Buy an S-Class Mercedes-Benz (sticker price: $103,895), and it comes with a seven-hundred-page owner's manual. Few people are going to read seven hundred pages, of course—which results in car owners making the very mistakes the manuals were designed to address. A few years ago, for instance, Subaru of America began to notice a rise in consumer complaints about the quality of its cars. When company executives looked into the situation, though, they found that the problem wasn't with the cars; it was with the owners. They didn't understand how the car worked . . . because they hadn't read the manual! In fact, one out of five calls to Subaru's call center involved a question answered in the owner's manual.

> One out of five calls to Subaru's call center involved a question answered in the owner's manual.

Such ignorance would be comical if it were not on occasion so serious. For instance, despite decades of public service advertisements stressing the importance of car seats (they can cut the risk of a child's death from an auto accident by 71 percent), most car seats are still installed incorrectly. One recent federal study pegged the rate of "critical misuse" at 73 percent—or nearly three out of every four. Why would parents habitually make such a mistake?

"They can't follow all the instructions," says Larry Decina, the lead investigator on the car seat study. "Go to your car's owner's manual tonight. Look how many pages it is for car seat instructions . . . It's like seventeen pages. You think people are reading that? Maybe Mommy looks at it. You think Daddy does?"

It's no wonder, then, that given a task to accomplish, we put the instructions aside and set out from our own mental model of how things work (or ought to work). But our models, unlike those formed by true experts, often contain hidden flaws that lead us into error. For instance, our intuition about how things work is not always right. This is particularly true for things that move. Most of us still hold intuitive theories of physics that were common in the three centuries before Newton. A classic demonstration of this (and another good bar bet) involves the following test: A flying plane drops a bomb. Which way does the bomb fall?

Most people think the bomb either drops straight down or even trails behind the plane. And most people are wrong. The correct path is a forward arc, like so:

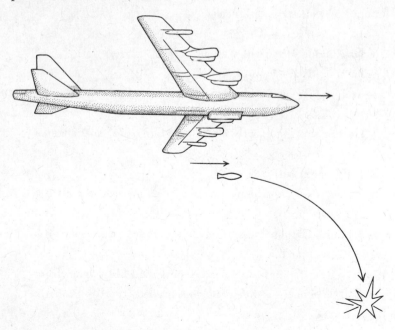

This error is so common (even among physics students) that it has been given a name: "the straight-down effect." It crops up in all sorts of everyday contexts. Take a situation seen on TV screens every weekend during football season: a running man drops a ball. Which way does the ball fall? It takes the same path as the bomb. But when a similar question was asked of sixth-grade students in Boston schools, only 3 percent got the answer right.

### One-Trick Ponies

Another problem with the bushwhack approach is that people tend to be one-trick ponies. If we learn to do something a certain way, we tend to stick with it. Psychologists refer to this mental brittleness as "functional fixity." This trait was famously demonstrated more than half a century ago through an ingenious experiment by Abraham and Edith Luchins. On its face, their experiment was simple: they just asked people to fill one of three jars with a certain amount of water. For instance:

Jar A holds 21 cups of water.
Jar B holds 127 cups of water.
And jar C holds 3 cups of water.

How do you get exactly 100 cups? Here's the cheat sheet:

1. Fill jar B with 127 cups.
2. Pour 21 of those cups into jar A; that leaves you with 106 cups in jar B.
3. Pour 3 of those cups from jar B into jar C; that leaves 103 cups in jar B.
4. Now, dump out the 3 cups in jar C and fill it again from B. When this is done, 100 cups are left in jar B. Problem solved.

Not so easy, huh? After repeating this experiment a few times, though, people got the hang of it. They usually discovered that this same pattern would work for all the problems they were given: from B, fill A, and then fill C twice.

But then the researchers changed things. They gave their subjects a second set of problems to solve. For these problems, the same pattern as before would work. But this time, there was a much simpler solution: just pour from jar A into jar C. For example, if jar A holds 14 cups, jar B holds 36 cups, and jar C holds 8 cups, how would you get just 6 cups? Answer: pour 8 cups from jar A into jar C.

The researchers repeated this experiment thousands of times. In all, the experiments yielded remarkably similar results: between 64 percent and 83 percent of the participants in both sets of problems continued to use the old, cumbersome method even when the new, simpler solution was available.

But here's the kicker: the Luchinses tried out the second set of problems on a fresh group of subjects who hadn't participated in the first round of problems. Nearly all of them figured out the simpler way. In fact, only between 1 percent and 5 percent of them failed to apply the simpler solution. The conclusion from these experiments was obvious: People in the initial experiments had become so set in their ways that they were blinded to the newer, simpler solution. But to those who came to the problem fresh, the simpler solution was obvious.

> People become so set in their ways that they are blinded to newer, simpler solutions.

## Thinking Outside the Box

In short, even though we often prefer to bushwhack our way toward a solution, most of us tend not to be very creative when it comes to solving the problem at hand, especially if we have already learned an approach that works. This tendency holds true even when we are presented with relatively simple (though novel) problems. Perhaps the most famous example—and a good one to try at home with friends, especially if you don't mind getting your walls marked up—is the candle problem. Here's how it works: give an unsuspecting friend three things—a book of matches, a box of tacks, and a candle. The task: attach the candle to the wall.

People will usually try to nail the candle directly to the wall with the tacks, which doesn't work, because the candle is too thick. Or they will try to melt the candle onto the wall. Very few of them think of using the box as a candleholder and then tacking the box to the wall. Instead, they just think of the box as a container for tacks. They do not, as the saying goes, think *outside* the box.

## Chapter 12

# We Don't Constrain Ourselves

One *way to reduce* errors is by introducing constraints. What are constraints? Essentially, they're simple mental aids that keep us on the right track by limiting our alternatives. I like to think of them as "bumpers" that nudge us back on course. But another way to think of them is as error blockers.

Constraints come in all shapes and sizes. Some are physical and some are not; they can be colors or smells, sights or sounds. The color red, for instance, works well as a constraint because we equate the color red with the action stop. Certain aspects of music also work well as constraints. As we saw with "The Star-Spangled Banner," a song's melody can serve as a constraint against forgetting. This helps explain why jingles stick with us long after we've seen the commercial and why, even as adults, we recite our ABCs to the same melody we learned as children.

The presence of constraints is not always obvious. Indeed, in well-designed products you probably won't even notice they're there. Take a pair of scissors, for instance. Can you spot the constraints? The constraints are in the size and shape of the holes: one is round and can fit the thumb; the other is elongated and affords room for the other fingers. The holes limit the way the scissors can be held and thus constrain the way the scissors may be used—or misused.

A concept that is closely related to constraints is one of affordances. Affordances are clues to how a thing can be used. Like constraints, affordances may appear in many forms: the shape, texture, or size of a thing may provide clues to its usage. A ball, for instance, affords bouncing or throwing. A knob affords turning. Slots afford the insertion of things. When we encounter some new object, its affordances help us answer basic questions like "How does it work?" and "What can it be used for?"

A number of famously well-designed products feature strong constraints and affordances. Legos are a good example. The cylinders and holes on Lego blocks function as natural physical constraints; they make it nearly impossible to connect Lego pieces in the wrong way. Velcro is another good one. It's immediately obvious to us what pieces of Velcro afford: sticking together! The same can be said for a variety of other products, from Frisbees to hula hoops to Post-it notes. Their use is so obvious, and grasped so immediately, that no directions are necessary; more important, operation of these items is virtually error-free.

Granted, these items may not strike you as sophisticated feats of engineering; but a great many products we use each day nonetheless lack their functional design. Have you ever walked up to a glass door in an office building and not known whether you should push it or pull it? That's because the door lacks proper affordances. (Does it afford pushing or pulling?) In such a case the resulting error is probably small—you may experience momentary confusion and perhaps a bit of embarrassment, but that's about all.

### Making the Same Mistake—Twice

But the lack of effective constraints can lead to errors that are far more serious. One of them occurred in November 2007 and nearly killed a pair of newborn twins. The twins, Thomas Boone and Zoe

Grace Quaid, were born prematurely on November 8. A few days later they came home from the hospital with their parents, the actor Dennis Quaid and his wife, Kimberly. But soon, the twins developed an infection. On November 17, a pediatrician told the Quaids to take the children to one of the nation's most well-known hospitals, Cedars-Sinai, in Los Angeles. This is where the mistake occurred.

The following day, November 18, the twins were injected twice with a massive overdose of the blood thinner heparin, which is often used to flush the tubes used to deliver intravenous medications to infants. At 11:30 a.m. and again at 5:30 p.m., nurses at the hospital mistakenly administered heparin with a concentration of ten thousand units per milliliter—instead of the standard dose of ten units per milliliter.

How could such a mistake happen—*twice?*

The easy answer is "human error." Indeed, Cedars-Sinai acknowledged that its staff made "a preventable error" by, among other things, not thoroughly checking the labels of the drugs that were being injected.

But a better answer is that the vials of heparin lacked strong constraints. The ten-unit vial and the ten-thousand-unit vial were easily confused with each other; both were of the same size and shape, and both had blue labels—albeit one dark blue and the other light blue. The potential to mistake the two was not only obvious but well-known. Between 2001 and 2006, more than *sixteen thousand* heparin errors were blamed on incorrect dosing. And in 2006, three babies at an Indianapolis hospital died after receiving a heparin overdose nearly identical to the one received by the Quaid children. Following the deaths of the Indiana children, Baxter Healthcare, which makes heparin, issued a warning. It cited "the potential for life-threatening medication errors involving two heparin products."

Baxter eventually changed the color of the label on the stronger dose from blue to red. But it did not make this change until October 2007—just one month before the Quaid children were overdosed. Even after the change, however, it never recalled the old vials, including the ones used at Cedars-Sinai. The resulting overdose left the Quaid twins' blood too thin to clot—a condition that was discovered only after nurses noticed that Zoe was oozing blood from an intravenous site on her arm and a spot on her heel. The babies were later given a drug to reverse the effects of heparin and restore the normal clotting abilities of their blood; after eleven days in the intensive care unit, the twins were released from the hospital and appear to have made a full recovery.

### What's in a Name?

Mix-ups like this are far more common than most of us know. In early 2008, just a few months after the overdosing of the Quaid twins made the news, the U.S. Food and Drug Administration issued a warning about two other drugs that could easily be mistaken for each other—not because their packaging was similar but because their *names* were. The drugs are edetate disodium and edetate calcium disodium. One is used to treat patients with high levels of calcium in the blood; the other is used to treat lead poisoning. Both are commonly referred to as EDTA.

Like heparin, the problems with EDTA were well-known. The FDA said it had received almost a dozen reports of deaths among children and adults over the past thirty years, including children being treated for lead poisoning who were given the wrong drug. In 2006, a separate federal agency, the Centers for Disease Control and Prevention, detailed some of the deaths in a report; one of them involved a child with autism who underwent treatment with one of the drugs to remove mercury from the blood.

In the United States, the FDA has the authority to approve the names of drugs. But the names of prescription drugs often give no clues (or affordances) as to what they are supposed to do. Many are barely pronounceable, let alone decipherable.

Zofran, anyone?

Zosyn?

Xigris?

Cubicin?

These are all real names of real drugs. Zofran, for instance, is an antinausea drug. Zosyn is an antibiotic, as is Cubicin. Xigris treats sepsis. But you would never know this by the names of the drugs alone; they could just as easily be those of a board game or a Greek god.

One could argue that the names don't need to mean anything because the drugs are supposed to be administered by people (like doctors and nurses) who are highly trained and should know what they are supposed to be used for. But why take the chance that a nurse pulling a double shift will keep the names straight?

### How Pilots Fly Straight

Contrast, if you will, the names doctors find on drug labels with the names that pilots find on their charts. To arrive in Nashville, for instance, pilots will encounter some PICKN and GRNIN and often a pass through HEHAW. That's because PICKN, GRNIN, and HEHAW are fixed points in the sky that pilots use when they are flying into Nashville International Airport in Tennessee.

Throughout the world, aviation authorities establish set routes to guide planes. They label key navigational points with unique identifiers, usually five-letter codes, called fixes. In the United States this job falls to the Federal Aviation Administration. Unlike the FDA, the FAA has actually chosen names that mean something; typically, they reflect well-known characteristics of the cities below. Passing over San Antonio? The fix is ALAMO. Headed to Orlando? You'll find MICKI, MINEE, and GOOFY. Some names are even mildly risqué: West Coast pilots who are headed east may notice BUXOM, which is in Oregon, followed by JUGGS, which is in Idaho.

> Passing over San Antonio? The fix is ALAMO. Headed to Orlando? Say hello to MICKI, MINEE, and GOOFY.

It wasn't always this way. The FAA had long used meaningless combinations of letters, some based on Morse code, to identify the fixes. But it began using five-letter pronounceable names in 1976 to improve safety by making it easier for pilots to remember instructions and avoid flying the wrong route.

"We try to select something that will be easily recognizable," said Nancy Kalinowski, an FAA official whose department oversees the name selection. "Any time you create uncertainty in the aviation world, there could be trouble."

### Lessons for Error-Free Design

This, of course, is true in the non-aviation world as well: uncertainty often equals error. We saw this with the confusion over the vials of heparin administered to the Quaid children. Their near tragedy offers at least two lessons in the understanding of why we make mistakes: one for the design of products, the other for the attribution of errors. Many of the everyday products we use are needlessly complex, prompting the uncertainty that leads to error. Immediate examples that come to mind are VCRs (still have one?) and digital watches. Figuring out how to use them is maddeningly difficult.

VCRs are (mostly) gone, but we must still grapple with devices that are needlessly complex. In 2001, for instance, BMW, a leading maker of luxury cars, outfitted its 7 Series automobile with its vaunted iDrive system. Essentially the system was a knob that controlled *more than seven hundred separate functions,* from climate and navigation systems to programmable settings for locks and lights.

The iDrive system, predictably, prompted an outcry from BMW fans. ("iDrive?" asked a headline in *Road & Track* magazine. "No, you drive while I fiddle with the controller.") They panned the device as nonintuitive and complicated. Simple operations often took several steps, and drivers had to take their eyes off the road to monitor the screen. After several years of criticism, BMW bowed to the complaints. In 2004 it said that it would offer a "simplified version" of iDrive on new models and make it an option on its redesigned 3 Series.

"Blessedly," wrote one reviewer, "the 3 Series can be ordered without the iDrive systems interface and its dreaded rotary control knob."

The lesson here should be obvious: simplify where you can, and build in constraints to block errors. This is what Baxter ultimately did with heparin. First, it made it easier to distinguish high-dose and low-dose vials by changing the color of the label on high-dose heparin from blue to red. It also added snap-off caps to the vials so that nurses must take an extra step when opening them. In addition, it enlarged the font size on the label to make it easier to read. And it stamped "Not for Lock Flush," referring to the low-dose flushing product, on the high-dose vials.

> **Strip out complexity and build in constraints.**

## Looking for Root Causes

As we saw from the overdose of the Quaid children, mistakes attributed to human error often have deeper roots elsewhere. This is one reason why we so often fail to learn from our mistakes: we haven't understood their root causes.

Ferreting out the root cause is not al-

> **The misattribution of errors is one reason we fail to learn from our mistakes: we haven't understood their root causes.**

ways easy. Like the leak that appears on our living room ceiling, the source of the problem may lie far from the spot where we notice it. In the case of human error, root cause analysis requires a deep understanding of human motivation. As we have seen in previous chapters, we believe we will act in one way, but often act in another—even in ways that would appear to be against our own self-interest. Even worse, many of us don't know when we're being biased. Our judgments may be distorted by overconfidence or by hindsight or by any of the other tendencies we've talked about.

People who are serious about eliminating errors would do well to keep this in mind. If you know, for instance, that the people who use your product are prone to overconfidence, it may make sense to design the product—or the way the product is used—in a way that counters some of that overconfidence. This is what happened at Methodist Hospital, the Indianapolis health care facility where three babies died from a heparin overdose. After the tragedy, Methodist replaced the ten-thousand-unit heparin vial with a heparin-filled syringe that cannot be confused with the smaller dose. In addition, now *two* health care workers must look at a dose of heparin before it is administered to a newborn.

A number of high-risk occupations have adopted this sort of belt-and-suspenders approach. Since 2004, for instance, doctors have been required to mark surgical sites on their patients' skin with a Sharpie or other kind of permanent marker to avoid cutting in the wrong place. That way, they don't have to rely on their memories. Pilots have long done something similar. Rather than trying to commit important details to memory ("I did check the flaps before takeoff, didn't I?"), pilots use a simple mental aid: a checklist.

> Ever wondered how bartenders remember complicated drink orders? The key is in the drinking glass.

Even bartenders (not exactly a high-risk occupation, I'll grant you) learn not to rely exclusively on their memories if they want to keep their drink orders straight.

Ever wondered, for instance, how bartenders remember complicated drink orders? ("Gimme a mai tai, a Rob Roy, a fuzzy navel,
and three mojitos.") Researchers at the City University of New
York found that the key is in the drinking glasses. The glasses, if
placed on the bar while the order is given, strongly limit the range
of possible drinks that could fill the order. Only a highball goes into
a highball glass, for instance, and only champagne goes into a
champagne flute. In other words, the glasses act as a poor man's
constraint, restricting what can be poured into them.

### Knowing Where to Look

Identifying the source of an error also requires knowing where and
how to look. After something goes wrong, we tend to look *down*—
that is, we look for the last person involved in the chain of events
and blame him or her for the outcome. But this approach, satisfy-
ing though it may be, usually doesn't stop an error from being
repeated—which is why the Quaid twins were overdosed not once
but twice —and by separate people. If multiple people make the
same mistake, then that should tell us something about the nature
of the mistake being made: its cause probably isn't individual but
systemic. And systemic errors have their roots at a level *above* the
individual. Which is why, when looking for the source of errors, it
pays to look up, not down.

Many of the errors we make are by-products of the culture of
the places where we work. Some places, for instance, tolerate
goof-ups more than others. Among those known for their lack of
tolerance is the U.S. Navy. Organizations like the Navy that perform successfully under very challenging conditions, with very
low levels of failure, are termed high-reliability organizations, or
HROs. Experts generally consider naval aviation to be a classic
HRO industry in that it achieves very low rates of catastrophic
failure while providing continuous service under hazardous conditions. The rate of class A accidents—which involve a fatality or

damage greater than $1 million—in U.S. naval aviation since 1999 has been approximately 1.5 per 100,000 hours flown. That is down from approximately 50 per 100,000 hours flown in the 1950s.

A number of organizations like to say that they, too, are HROs. (How many of us have worked at jobs with placards that proclaim, "Safety is No. 1"?) Among the top contenders for this title are hospitals. Hospitals routinely claim to be in the safety business. But which is the more safety conscious: hospitals or aircraft carriers?

To find out, Dr. David Gaba, a well-known researcher in the field of patient safety, and colleagues from Stanford University and the Naval Postgraduate School in Monterey, California, submitted surveys to 15 hospitals and 226 squadrons of naval aviators. The surveys contained twenty-three common questions. The wording of these questions was not always identical. For instance, the surveys given to pilots contained the phrase "command leadership," whereas hospital surveys would use the term "senior management" or "my department." But in the main, they were quite similar. Gaba and his colleagues then classified the responses into certain categories. For each question, for instance, a "problematic response" was defined as one that suggested an absence of safety. What they found was eye-opening.

> Overall, the problematic response rate for pilots was relatively small—just 5.6 percent; but for doctors and nurses, it was more than three times as high: 17.5 percent.

Overall, the problematic response rate for pilots was small—just 5.6 percent; but for doctors and nurses, it was more than three times higher: 17.5 percent. And on some questions, the problematic response among hospital workers was higher still: up to twelve times greater than that among aviators.

## Attitude and Errors

Many explanations could account for this gap. Unlike the military, for instance, most hospitals aren't centralized organizations; there's no admiral who can hire and fire on the spot; indeed, physicians' groups often act autonomously within a hospital. Gaba concluded that while military commanders had made great strides in instilling a culture of safety, hospital leaders had not. Their "avowed commitment to safety," he wrote, "has not translated sufficiently into a climate in which safety and organizational processes aimed at safety are valued uniformly."

Another factor that helps explain the difference in attitudes between doctors and pilots may be the nature of the work. Unlike pilots, doctors don't go down with their planes. This difference gives pilots a powerful incentive to eliminate errors: the lives they save may be their own.

Remember how the Navy's accident rate had improved, from 50 class A accidents per 100,000 hours flown in the 1950s to 1.5 per 100,000 hours now? A similar trend has occurred among civilian aviation: in the last ten years fatal crashes of airplanes in the United States have declined 65 percent, to one for every 4.5 million departures in 2007 from one in nearly 2 million departures in 1997. The rate is so low that in 2007, the outgoing administrator of the FAA said, "This is the golden age of safety, the safest period, in the safest mode, in the history of the world."

> Studies of autopsies have shown that doctors seriously misdiagnose fatal illnesses about 20 percent of the time. That's one out of five.

Now take a look at the record for doctors. There are many ways to do this, of course, and not everyone will agree on what "record" to look at. But consider this: studies of autopsies have shown that doctors seriously misdiagnose fatal illnesses about 20 percent of the time. That's one out of five. So millions of people are treated for the wrong disease. But the really shocking thing is this: the rate of misdiagnosis has not really changed since the 1930s. "No improvement!" was how one article in *JAMA: The Journal of the American Medical Association* summarized the findings.

Again, there can be many reasons for this; entire books have been written on the appalling rates of medical error. (In the United States, between forty-four thousand and ninety-eight thousand patients are thought to die each year from preventable errors—which would make medical errors the eighth most common cause of death.) And the human body is certainly much more complex than an airplane is.

But, again, consider just one: attitude. Attitudes inside operating rooms differ quite a bit from attitudes inside cockpits. ORs are typically hierarchical places, with the surgeon on top; cockpits are not. Flight crews, which typically consist of a captain, first officer, and second officer, are encouraged to speak up if they see something amiss, no matter their rank; when it comes to pointing out potential errors, everyone is considered equal.

> One survey asked whether junior staff members should be free to question the decisions of senior staff members. Ninety-seven percent of pilots said yes. But when surgeons were asked, only 55 percent of them agreed.

This difference in attitude was revealed in a recent survey given to tens of thousands of pilots and doctors and other health care workers in the United States, Europe, and Israel. The survey asked, among other things, whether junior staff members should be free to question decisions made by senior staff members.

When airline pilots were asked, nearly all of them—97 percent—said yes.

But when surgeons were asked, only 55 percent of them said yes.*

This difference is attributable, in part, to a modern airline safety science called Crew Resource Management. CRM, which is taught to flight crews around the world, grew out of a crash in 1978 near Portland, Oregon. The cause of the crash wasn't complicated: six miles short of the airport, a United Airlines DC-8 simply ran out of fuel. The flight engineer knew the plane was running out of gas, but didn't tell the captain until it was too late. Of the 189 passengers and crew on board, 10 were killed. The incident led to pilot training in communications and cooperation, teaching crews how to work together. Today, CRM is an aviation industry standard, taught to flight crews around the world.

### A Mutilating Surgery

It is not always easy to identify how something as nebulous as one's attitude or a team's inability to work together can lead to errors. But one chilling example was documented by Charles Vincent, a professor of clinical safety research at Imperial College London. The case, which Vincent reviewed at the request of the Agency for Healthcare Research and Quality, involved a thirty-three-year-old woman who had received a terrible diagnosis: invasive cancer of the vulva, the sensitive area outside of the vagina that includes the labia and clitoris. To eliminate the cancer, doctors proposed a radical surgery: removing half of the woman's vulva.

After the woman was placed under general anesthesia, a med-

---

*Pilots were also far more willing to acknowledge their limitations. One of the items in the survey posed the following statement: "Even when fatigued, I perform effectively during critical times." Seventy percent of surgeons agreed with the statement, but only 26 percent of pilots did. See Sexton, Thomas, and Helmreich (2000).

ical trainee at the hospital reviewed her chart and noted a critical detail: the biopsy had indicated that the cancer was from the *left* side of the vulva. But, just as the trainee prepared to cut into the left side of the vulva, the attending surgeon stopped him. He said the trainee should cut on the *right* side.

At first, the trainee balked. He told the surgeon that he had just reviewed the chart and that the positive biopsy hadn't come from the right side—it had come from the left. But the surgeon would not back down. He told the trainee that he himself had done the biopsy and remembered that it was taken from the right side. Faced with this insistence from his superior, the trainee did what the surgeon told him to do and removed the right side of the woman's vulva.

As is standard procedure, a sample of the removed tissue was sent to the hospital's lab to be checked for signs of the cancer. But when the pathologist examined it, he was startled; it was cancer-free. Armed with this information, the trainee went back to the surgeon and told him that there had been a mistake; they had removed the wrong half of the vulva. But the surgeon denied that any error had been made; in fact, he insisted that the original biopsies had been mislabeled.*

When she returned for a routine follow-up, the surgeon performed a biopsy of the left side of her vulva—which, of course, showed cancer. Shortly thereafter, the woman was operated on again; this time, the rest of her vulva was removed.

Though anecdotal, the incident reviewed by Professor Vincent is not isolated. Wrong-site surgery continues to afflict untold numbers of patients each year. One study suggests that it is rela-

---

*Note the similarity here between the surgeon's attitude and those of the "experts" we saw in an earlier chapter: even in the face of evidence to the contrary, both tried to convince themselves they were right.

tively rare, occurring once in every 113,000 surgeries. But the full extent of the problem is unknown, and many surveys suggest that it is likely to be underreported. One recent survey, for instance, asked hand surgeons about operating on the wrong place; 20 percent of them revealed that they had operated on the wrong site at least once in their careers.

> In one survey, 20 percent of hand surgeons revealed that they had operated on the wrong site at least once in their careers.

*Chapter 13*

# The Grass *Does* Look Greener

In *the summer* of 2000, the newlyweds Pete and Jennifer Marino made a big decision. They decided to move from their home in Wisconsin, where they had lived most of their lives, to a place they had never lived before: California.

"I was totally fired up to be in warm weather," says Jen, sitting on a sofa next to Pete, her brunet ponytail poking out from underneath a baseball cap. Like Pete, she was born and raised in Wisconsin and was eager to try something new. Pete, who had been working for the Miller Brewing Company, had been accepted into the business school at the University of California, Los Angeles. He had also been accepted to business schools at the University of Michigan and at Duke. But Michigan and North Carolina, he said, held none of the mystique of Southern California.

"So we decided to go to L.A."

They ended up renting a place sight unseen in Westwood and plunged into L.A. life. They took a helicopter tour over the city. They drove by the Hollywood sign. Went to the beaches. Went to the mountains. Went to Lakers games.

Gawked.

"We did have a lot of celebrity sightings," says Jen. "I'm not above that. That was fun!"

"I mean, we would go to church every Sunday, and Mark Wahlberg sat right in front of us," says Pete. "Good Shepherd Church. In Beverly Hills. The church Frank Sinatra had his funeral at."

In 2003, Jen became pregnant with their first child, Max. As is often the case for new parents, life changed. At first, it was the little things Pete noticed.

"For instance," he says, "before Max was born, we had people asking us, 'Well, what waiting lists for schools are you on?' "

"For *pre*school," says Jen.

"Before he was born!" says Pete.

"And volunteering: 'Haven't you started volunteering at the school you want your child to go to?' "

"And they were like, 'Well, right after he's born, make sure you call and you're on the list,' " says Pete. "That whole thing—that just sits weird with us."

### "So Here's a Funny Story for You About L.A."

And then there was the neighborhood. Guys Pete knew were living in expensive areas like the Pacific Palisades, where an eleven-hundred-square-foot home fetched $1.7 million. But with Pete fresh out of business school, neighborhoods like that were out of the question. So he and Jen bought a more modest home in West Los Angeles.

"We had to put the alarm on every night," says Pete. "There were police helicopters with their lights on going throughout our neighborhood all the time—I mean, at least five nights a week. My car got broken into. Our nanny's car got stolen in front of our house."

And they saw things that they just wouldn't see in Milwaukee.

"You would drive down the street and see a mother and daughter wearing the same clothes," says Pete.

"With their Starbucks cappuccino," says Jen.

"And you'd hear stories about kids getting cosmetic surgery, thirteen, fourteen years old—"

"Birthday presents," says Jen.

And that didn't sit well with them, either. After a while, a lot of things didn't sit well with them. Before long, Pete and Jen began to feel a creeping sense of estrangement from their new hometown.

"So here's a funny story for you about L.A.," Pete tells me. "So Max is getting baptized at this church, right? And we have to go to this baptism class. And that stuff drives me crazy. So I'm whining and bitching about it. And Jen says, 'You gotta go, all right?' So I got there, and there's, like, not that many people there. A handful of couples. I see a woman and her husband: 'Hey, how are you?' Charisma-and-something. Okay. He's a nice guy, she's a nice girl— cute, whatever. A couple months later—"

"No, you called it right then," says Jen, elbowing Pete in the shoulder.

"I knew who she was right away," says Pete. "She was Charisma Carpenter, and she's in *Buffy the Vampire Slayer* TV show, okay?"

"No, that was not the reference you gave me—"

"It was after that," Pete says. "Anyway, so then I'm back in Milwaukee, eight months later, in the media director's office at Miller Brewing Company. And because Miller advertises in *Playboy,* this guy has got, you know, *Playboy* magazines. And I'm looking on his desk, and I'm like, 'Gimme that thing!' And on the cover of *Playboy* is Charisma Carpenter. Who Max got baptized with their kid! With, you know, a full spread in there! That's an L.A. story. That's the kind of shit that happens out there."

Somewhere between the property values and the moral values, Pete and Jen decided California was no place to raise a family. So in 2004, four years after they moved from the Midwest, they decided to move back. It was not an easy decision.

"I cried my eyes out in the two weeks before we left," says Jen. But leave they did. And so did a couple million other people. At last annual count, more than 2.2 million Californians packed up and moved to another state, according to the Census Bureau. But nearly 1.5 million people moved right back in, drawn by the sunshine and the beaches and the same mystique that attracted the Marinos. Every year the Harris Poll asks Americans to name the state where they would most like to live. And every year for the last five years the answer has been the same: California.

> Every year the Harris Poll asks Americans to name the state where they would most like to live. And every year for the last five years the answer has been the same: California.

## Predicting How We (and Others) Will Feel

"Everyone should be happier in California, right?" asks David Schkade, who is himself a recent émigré to the Golden State, having left his home state of Texas to become a professor at the University of California, San Diego. As a psychologist who teaches at the university's business school, Schkade is interested not only in why people think they will be happier in California but, more broadly, in how well people are able to predict how they will feel about things in the future—whether it's moving to California or retiring to a life of leisure.

On its face, this is a subject that shouldn't need much studying. Most of us, after all, are pretty good at predicting how we will feel in the future. If our house burns down or our dog dies, we know we will feel sad; and if we get a promotion or have kids, we will probably feel glad (at least until the tuition bills come around).

But from afar, life often looks different from how it does close-up. Many times, we don't know how we will feel about something until the moment arrives. For instance, if you're young and healthy, you probably don't want any exceptional measures taken to prolong your life once you're old and drifting away on that ice floe to oblivion. In fact, this is what researchers have found. When they asked doctors and other healthy people if they would accept a grueling course of chemotherapy if it would extend their lives by three months, they got very few takers.

> Would you accept a grueling course of chemotherapy if it would extend your life by three months? Most people said they wouldn't. But 42 percent of current cancer patients said they would.

No radiotherapists said they would do it.

Only 6 percent of oncologists said they would do it.

And only 10 percent of healthy people said they would do it.

Ah, but none of them was dying (at least not right away). When researchers asked the same question of people who *were* facing an imminent risk of death—current cancer patients— they got a much different answer: 42 percent said they would do it. Another study found that 58 percent of patients with serious illnesses said that when death was near they would want treatment—*even if it prolonged life by just one week.*

Life and death undoubtedly present a special case. But people have also been shown to systematically mispredict how they will feel in the future after experiencing a number of lesser (though still major) life events, like moving or breaking up with a romantic partner or even failing to get a promotion. One long-term study followed high school students who smoked cigarettes. Only 15 percent of the students who were light smokers (less than a cigarette a day) predicted that they would still be smoking in five years. But five years later, 43 percent of them were still smoking.*

## Why Gift Cards Are a Bad Idea

We make this kind of error often enough that it can be—and is—reliably used against us. Rebate coupons, for instance, are a great way to get people to buy things; studies have shown that our purchasing decisions are often swayed by rebate offers. But here's the catch: many of us never use them. By one estimate, 40 percent of rebate coupons are never redeemed. If you've ever wondered why merchants offer rebate coupons instead of just cutting prices by an equivalent amount, that's why: they know that consumers (like high school smokers) will mispredict how they will behave in the future.

This misprediction is why gift cards are such a bad idea for you—but a really terrific idea for the companies that issue them. Gift cards have become tremendously popular in recent years and are now the No. 1 gift choice in America, with two-thirds of consumers saying they plan to buy one. There's only one problem: the people we give them to don't use them. On average, U.S. consumers have between three and four unused gift cards apiece lying around in purses and dresser drawers. And these unused cards add up: Americans lose about $8 billion annually by not redeeming gift cards. But this is money in the bank for the companies that issue them. In 2008, for instance, Limited Brands, owner of the Victoria's Secret chain of lingerie stores, reported a quarterly pretax gain of $47.8 million, or eight cents a share, from unused gift cards. And it

*Heavier smokers weren't much different. Only 32 percent of those who smoked at least a pack a day in high school said they would be smoking in five years; five years later, 70 percent of them were still pack-a-day smokers.

isn't alone. Big retailers like Target, Best Buy, and Home Depot also make a killing on unused cards.

Researchers call this type of error a projection bias, and it pervades even the most intimate of decisions. Spouses, as you may know from your own personal experience, often aren't very good at predicting what types of presents their mates would like for Christmas or other occasions. What often happens is that spouses first decide whether *they* would like the present in question, then adjust from there to decide whether you would like it, too. ("Really, honey, I just love that thong . . .") Consequently, many retailers, especially high-end boutiques, have tried to remove the guesswork from giving gifts to the people we think we know so well. For instance, the online retailer Net-a-Porter.com, which sells women's designer clothes and accessories, e-mails videos to boyfriends and husbands telling them what's on their significant other's wish list. The videos feature a flirtatious blond "Santa's helper" who addresses the recipient with preselected pet names like "Honey Bunny."

"Let's face it," she says, "if she's happy, you're happy."

And if there's one thing we all want to be, it's happy. But in trying to figure out the happiness puzzle, we often focus on the wrong things. Consider the unusual case of people who have undergone a life-changing medical procedure, such as a colostomy. Who's happier: those with a permanent colostomy or those who have a chance of reversing it?

The Carnegie Mellon professor George Loewenstein and colleagues posed that question while researching patients at a Michigan medical center who had undergone colostomies. Half of the patients faced the possibility of having their colostomies reversed at some point in the future; for the other half, the operation was permanent: for the rest of their lives they would have to defecate into a bag attached to their bodies. Over a six-month period the patients were asked to rate their overall life satisfaction.

If you're like most people, you probably think those with the potentially reversible colostomies would be happier (I did). After all, most of us dread finality; we much prefer changeable outcomes. We like escape hatches and will often pay more for the privilege of changing our minds. That's one of the reasons why we have adjustable-rate mortgages and prenuptial agreements, and shop at retailers like Costco that offer liberal return policies (so we can return those Christmas presents from our spouses). If we make bad decisions, we want to be able to get out of them.

But it turns out that those with the permanent colostomies were happier. Throughout the six-month period the researchers found that those with permanent colostomies improved rapidly. But those who could ultimately reverse them remained relatively unsatisfied. Why? Loewenstein's conclusion is that "hope impedes adaptation." In other words, if you're stuck with something, you learn to live with it. And the sooner you learn to live with it, the happier you will be.

> "Hope impedes adaptation." In other words, if you're stuck with something, you learn to live with it.

This finding is consistent with a long line of psychological research going back to Freud. Upon being stuck with a decision, we suddenly feel it's not so bad. Voters, for instance, have been shown to recognize the strengths of a candidate they opposed—once that candidate is elected. And high school students become acutely aware of a college's weaknesses—upon learning that it has rejected them. College students, likewise, come to appreciate how biased standardized tests are—after failing one. In other words, they adapt. But from afar, this is often not what we expect.

## Focusing on the Wrong Things

The power expectations have to shape our predictions for future happiness is one of the things that intrigues Schkade. When we consider the impact of any single factor on our well-being, says Schkade, we are prone to exaggerate its importance—a tendency he and his colleagues have labeled a "focusing illusion." This illusion, he says, is a hidden source of error in significant decisions we make about our lives.

In a series of experiments, Schkade, Daniel Kahneman, and others demonstrated that when faced with important life decisions, we often tend to focus on relatively minor factors—like the weather in California—and downplay the other factors that will make up the bulk of our daily lives there, like the length of the commute or the cost of living. As a result, we often think we'll be happier than we end up being. We get to the other side of the hill only to find the grass isn't as green as we thought.

As counterintuitive as it may sound, when it comes to personal happiness, there's a good deal of evidence that circumstances don't matter as much as we might think. Indeed, researchers have found that when it comes to well-being, neither social status, education, income, marital status, nor religious commitment accounts for more than about 3 percent of the variance in people's reported levels of well-being. Persons with disabilities ranging from quadriplegia to blindness consistently report surprisingly high levels of well-being. For individuals with extreme quadriplegia, for instance, 93 percent report being glad to be alive, and 84 percent consider their lives to be average or above average. Similar levels of happiness are reported even by those with severe, multiple handicaps. Most of the people in this study were unemployed, unmarried, and relatively uneducated. Yet 96 percent were

> Among people with extreme quadriplegia, 93 percent report being glad to be alive, and 84 percent consider their lives to be average or above average.

satisfied with their living arrangements, 82 percent with their social lives, and 76 percent with their education.

Taking a cue from this line of research, Kahneman and Schkade gave questionnaires to 119 students. The questionnaires asked the students for their opinions on paraplegics, including how often they thought paraplegics might experience bad and good moods. The results were unequivocal: people who knew paraplegics viewed them as happier than those who did not. More specifically, those who said they had never known a paraplegic thought paraplegics would have more bad moods than good, by a margin of 43 percent to 32 percent. But those people who had actually known a paraplegic as a friend or relative had just the opposite perception: 20 percent versus 53 percent. The message, says Schkade, is clear: the less you know about paraplegics, the worse off you think they are.

But what about other things? Do people exhibit the same biased way of thinking about other subjects they know little about? More particularly, say, do people think about moving to California in the same way that they think about paraplegics?

To find out, Schkade and his colleagues had professional survey firms recruit nearly two thousand students in the Midwest and in Southern California and paid them to participate in a one-hour session. During the session the students filled out a questionnaire that asked them to rate their overall satisfaction with life as well as various aspects of life, either for themselves or for someone like them in one of the two regions. Then the professors went to work, analyzing the answers. What they found may surprise you.

First, overall life satisfaction was the same in both regions. Midwesterners reported being just as happy as Californians, though they differed on certain aspects. Midwesterners, for instance, weren't as happy about the weather as Californians were. But overall, their satisfaction with life in general was the same.

But when the students were asked not about their own happiness but about whether someone like them would be happy, their assessments changed. Both groups expected Californians would be more satisfied than Midwesterners. Why? Schkade found that the students tended to focus on aspects of life that they themselves rated as relatively unimportant but that they *thought* would be important to somebody else. On the issues that mattered most to them—things like job prospects, economic opportunities, and crime—there was virtually no difference in assessments. Only when the decision making got down to the stuff that wasn't as important—like the weather—did differences emerge.

"When you're trying to make judgments about a complex subject, you tend to zero in on things that are easily observed and give too much attention to those," says Schkade. This is true, he adds, no matter the subject. He and his colleagues have found similar results when they look at how people make decisions in-

> We "tend to zero in on things that are easily observed and give too much attention to those."

volving issues like weight, age, and money. People with greater income, for instance, tend to spend more of their time engaged in stressful activities, like work or child care, and not in leisure activities, like watching TV. Yet, notes Schkade, they'll buy the big-screen plasma TV that they will seldom watch.

"Anything that you focus on," he says, "isn't as important as you think."

Which helps explain why millions of Californians, the Marinos among them, leave each year. In the end, said Pete, "living in California doesn't matter enough to me to take advantage of the weather or whatever else to make all those other calls in our lives."

In 2004, he and Jen put their house in West Los Angeles up for sale. At their first open house, more than two hundred people showed up. "We had people writing offers on the hoods of their cars," said Jen.

With the money from the sale they bought another house, this one in a suburb north of Chicago. Which is where, on a brilliant autumn afternoon, we talked over the din of a neighbor's leaf blower. Pete and Jen both still miss aspects of their life in California. She misses Christmas shopping in shorts; he misses being in the Pacific time zone and getting to watch football games at ten o'clock on a Sunday morning. "Stupid things like that," says Pete.

But Max has a new little sister, Sophie, and now this house, too, is up for sale. There are no offers on the hoods of cars, no two hundred people at their door. Still, the Marinos have already decided on their new destination. This time they're moving just a block away.

## Conclusion

So *what does* it all add up to? How can you make fewer mistakes? My advice:

Think small.

Each year in the United States some seven thousand people die—not from disease or accidents or acts of God (though these kill plenty); they die on account of doctors' sloppy handwriting. And remember the South African bank experiment? What prompted borrowers to borrow? Not low interest rates or liberal loan terms (though these were attractive). It was a woman's picture (which was also attractive). And what about the penalties against teams in the NHL and the NFL? The culprit wasn't the players or the coaches or even the referees: it was the color of the uniforms.

> "The tiniest little change in circumstance can have big impacts on people's behavior."

All little things. But the little things, as the song says, mean a lot. Cornell professor Tom Gilovich said as much when I asked him about the black uniforms experiment. "To my mind," he said, "it illustrated one of the most important findings from my field of psychology: the tiniest little change in circumstance can have big impacts on people's behavior."

At first, this insight may seem difficult to apply to our everyday lives because the connection between our circumstance and our behavior often is not apparent to us. We are blinded by the effects of habit and hubris and hobbled by a poor understanding of our own limitations. We don't see all that we observe, and yet we sometimes "see" things we don't know we've seen. We can judge someone's honesty or likability in the blink of an eye, for instance, yet not notice a change in the identity of a person carrying a door right by us. Our decisions, even about important subjects like life and death, can be swayed not only by the options presented to us but by the way in which the options are presented. We are heavily influenced by first impressions, even when we try hard not to be, and we have a strong aversion to reading and following directions, preferring instead to follow our own intuitions about how a thing can or should work.

There is an emerging consensus among some psychologists that human decision making operates on two levels—one more rational, one more visceral—and that these two constantly trade off, like a car's headlights blinking from high beam to low. Many of our mistakes appear to happen while we are operating in one condition but think we are operating in the other. We may think, for instance, that our decision to take out a loan was dictated by financial considerations alone, only to learn that we were influenced by that picture of a pretty woman. Or we may vote for a politician after listening to many hours of campaign coverage on TV, only to learn that we formed an opinion about his (or her) competence in the blink of an eye.

If this emerging view is accurate, it helps explain why some types of errorprone behaviors are so hard to eliminate: we think we're being rational when we're being visceral, and vice versa. When a mis-

> We often think we're being rational when we're being visceral, and vice versa. When a mistake does happen, we often end up blaming the wrong cause.

take does happen, we often end up blaming the wrong cause ("Why did I take out a loan with such a high interest rate? I must not have done the math right . . ."). We don't learn from experience, because we're not sure which experience to learn from.

To make matters worse, studies with children and adults show that a large percentage of people cannot tolerate mistakes. This is particularly true for people who believe that intelligence is fixed and cannot change; in their eyes, a mistake is a reflection of their intelligence, and there is a strong tendency not only to avoid making mistakes but, as a consequence, to avoid learning from them. For many of us, this tendency does not improve with age. As we get older, we become more vested in being right, both at home and at work. And when we are right (and let's not forget, we are right a lot of the time), we tend to attribute our rightness to our skill in whatever it is we're right about; but when we're wrong—that we attribute to chance.

## Keep Tabs on Your Dry Holes

Nevertheless, the good news is that some of our biggest biases are amenable to correction. Take overconfidence, for instance. As we saw in an earlier chapter, very few of us think we are average; we are, as researchers say, poorly calibrated. But calibration is nevertheless a skill that can be taught. As evidence, Paul Schoemaker, the Wharton professor who has studied managerial overconfidence, and his colleague J. Edward Russo cite the example of the global oil giant Royal Dutch Shell.

Shell had noticed that newly hired geologists were wrong much more often than their levels of confidence implied. For example, the geologists might estimate that a certain site had a 40 percent chance of containing oil. But when ten such wells were drilled, only one or two would produce. That cost Shell time and money.

What did Shell do? As odd as it sounds, it made its geologists more like weather forecasters. How? By designing a training program to help the geologists improve their calibration. As part of this training, the geologists received case histories that included many factors that could affect the presence of oil deposits. For each case, the geologists were asked not only to provide their best guesses but to put numbers to these guesses that were precise (just as weather forecasters use probability statements). Then, for each case, the geologists were told what had actually happened. (In other words, they were given direct feedback.)

The training, according to Schoemaker and Russo, "worked wonderfully." Now, when Shell geologists predict a 40 percent chance of producing oil, four out of ten times the company averages a hit.

> **Don't just keep track of the stocks you pick—keep track of the ones you thought about picking but didn't.**

You can conduct a poor man's version of this training at home. You may not pick oil wells, but you probably pick other things, like stocks. And if you do, don't just keep track of the stocks you pick—keep track of the ones you thought about picking but didn't. Write down the reasons for your decision at the time (this is important because creating a written record lets you fend off the rose-colored-glasses effect of hindsight bias). Then keep track of these stocks. How have they performed relative to the one you did pick? And why? Did your original reasons for not picking them pan out? Why not? In short, keep tabs on your dry holes. Studying their performance may help you improve your own.

### Think Negatively

Also, the next time you have a major decision to make, ask yourself: What could go wrong? This may strike you as needlessly pessimistic and even downright defeatist; since childhood most of us have been prodded to think positively, and with good reason. On dark days, a positive attitude may be all that keeps us from going off the deep end. But thinking positively has its limitations; among other things, it can blind us to the pitfalls that lay camouflaged inside our ideas.

Atul Gawande, a surgeon at Harvard Medical School, has called this approach "the power of negative thinking." In certain circumstances, he says, it can be critical to look for and even expect failure. He cites the experience of Walter Reed Army Medical Center. The death rate among soldiers was 25 percent in the first Persian Gulf war, but is only 10 percent now. Care itself did not change; medical personnel are actually stretched thinner now. But they tracked weekly data injuries and survival rates, and actively looked for failures and how to overcome them. One small example involved eye injuries. Rather than being content to simply treat the injuries, doctors asked a more unnerving question: Why had so many injuries occurred? It turns out that young soldiers weren't wearing their protective goggles—they were too ugly. So the military switched to cooler-looking ballistic eyewear. The soldiers wore the glasses more often, and eye injury rates dropped immediately.

The same approach works in business contexts, says Paul Schoemaker.

"If you get people to play devil's advocate with themselves— asking what the evidence against this is—overconfidence is pretty close to being eliminated."

So give it a try.

## Let Your Spouse Proofread

There are other things you can do to become less error prone. Many of them are small, and some may strike you as obvious. It helps, for instance, not to be so set in your ways. Habit is a great friend to us all, saving us time and much mental effort. But it can kill our ability to perceive novel situations. After a while we see only what we expect to see. We skim over things and see not details but patterns. Remember the Goldovsky error—which was caught not by the expert sight readers of music but by a novice? Keep that in mind as you try to weed out errors in your own life. If, for instance, you have a paper that you want someone to proofread, by all means give it to a respected colleague—but she will likely miss the same errors you miss. So also give it to your partner or even to your kids to read; they may find the errors that others miss.

It also helps to slow down. Multitasking is, for most of us, a mirage. There are strict limits to the number of things we can do at one time, and the more we do, the greater the chance for error. Remember Captain Robert Loft, the pilot who flew his jet into the ground? He became so distracted by his hunt for a $12 lightbulb that he forgot to fly the plane. We are similarly distracted by equally small things, like the GPS devices in our cars or the iPods in our ears. The use of these devices is so absorbing that we often lose track of our surroundings. Indeed, the risk of distraction is hazardous enough that in 2007, USA Track & Field, the national governing body for running, banned the use of headphones and portable audio players like iPods at its official races. Many times, runners were so zoned out that they couldn't hear warnings and collided with other runners.

It helps, too, to beware the anecdote. Remember the Nutri-System example? How did the diet company's executives get people to try its product? They didn't use statistics; they used anecdotes. When making decisions, we often give vivid information—like diet testimonials—more credence than it deserves. As a consequence, we frequently make a bad decision. Indeed, the power of

> When making decisions, we often give vivid information—like diet testimonials—more credence than it deserves.

anecdotes to lead us astray is so strong that an influential CIA study advises intelligence analysts to avoid them. Analysts, it concluded, "should give little weight to anecdotal and personal case histories," unless they are known to be typical, "and perhaps no weight at all if aggregate data based on a more valid sample can be obtained." That's good advice. So ask for averages, not testimonials.

### Get Some Sleep

Another often-overlooked cause of error is a lack of sleep. Sleepy people make mistakes, which we all know. But there are staggering numbers of sleep-deprived people out there (you may even be one of them). At last count, forty-two million prescriptions for sleeping pills were filled in the United States; that's about one for every seven Americans, a number that has increased 60 percent over the last five years. All this pill popping brings about its own errors, of course. People are so sedated that the federal government has begun to warn of a new peril: sleep-driving, which occurs when people drive while under the influence of sleeping pills. They also sleep-eat; people have reported ingesting buttered cigarettes or waking up gasping for breath with a mouthful of peanut butter, a particular sleep-eating favorite.

What many of us do not realize, though, is that the lack of sleep affects not only our physical and mental abilities; it affects our mood. Even moderate sleep deprivation, for instance, can cause brain impairment equivalent to driving drunk. And with increasing fatigue, as with increasing intoxication, people demonstrate a greater willingness to take risks—which is probably not what you want when those people are behind the wheel, or wielding a scalpel, or doing any of a thousand other jobs. Yet this is exactly what happens. Between 2003 and 2007, for instance, there were at least half a dozen cases in which pilots in the United States fell asleep—mid-flight! In one case, the pilot *and* the copilot fell asleep while descending toward Dulles International Airport near Washington, D.C. In another, Frontier Airlines acknowledged that two of its pilots fell asleep on a 2004 red-eye flight from Baltimore to Denver. Fortunately, one of the pilots woke up after "frantic calls" from a controller.

Clearly, some pilots need more rest. But what approach have

> With increasing fatigue, people demonstrate a greater willingness to take risks.

airlines taken? In 2005, thousands of passengers on JetBlue Airways became unwitting participants in an unusual experiment to test the limits of pilot fatigue. Without seeking approval from the Federal Aviation Administration headquarters, consultants for the airline outfitted a small number of pilots with devices to measure alertness. Then they assigned the flight crew to work longer hours than normal—ten or eleven versus the eight hours the federal government allows. JetBlue hoped its experiment, which covered more than fifty flights, would show that pilots could safely fly far longer without exhibiting ill effects from fatigue. And JetBlue isn't alone in trying to squeeze more work out of its pilots. In 2007 two of the nation's largest airline companies—Continental and American—balked at providing extra rest periods and other safety measures for pilots on international flights.*

And it's not just pilots, either. Young doctors face similar work pressures. Medical residents are routinely scheduled to work shifts that last twenty-four hours or more, even though these sleep-deprived doctors are at high risk of making medical mistakes that can hurt or kill patients. By one estimate, about 100,000 residents in the United States work such extended shifts. Charles Czeisler of the Harvard Medical School and his colleagues studied more than twenty-seven hundred first-year medical residents. They found that when residents reported working five marathon shifts in a

> When medical residents reported working five marathon shifts in a single month, their risk of making a fatigue-related mistake that harmed a patient soared by 700 percent.

---

*Air-traffic controllers face similar pressures. In 2007, the National Transportation Safety Board concluded that fatigued controllers contributed to not one but four airplane "mishaps" in recent years. One case cited by the NTSB involved the crash of a Comair jet in Kentucky in 2006. The sole air-traffic controller on duty that morning reported that his only sleep between shifts had been a two-hour nap.

single month, their risk of making a fatigue-related mistake that harmed a patient soared by 700 percent. The risk of making a mistake that resulted in patients' deaths rose by "only" 300 percent. The residents in Czeisler's study reported making 156 fatigue-related errors that injured patients. Thirty-one of them actually led to patients' deaths.

It's not hard to see where all the lack of sleep leads. It is, as the saying goes, an accident waiting to happen—except that when it does happen, there will be nothing accidental about it.

## How Happiness Helps

It also helps to be happy. In surveys, most people report that they are happy, which is fortunate because there is a correlation between happiness and certain kinds of errors. Happy people tend to be more creative and less prone to the errors induced by habit. In ways that are not fully understood, good feelings increase the tendency to combine material in new ways and see relatedness between things. One of the clearest ways to demonstrate this involves the ability to solve the candle-and-box-of-tacks problem we discussed earlier. People who are happy have been shown to do much better at this than people who aren't.

> Happy people tend to be more creative problem solvers. They also make decisions more quickly, with less back-and-forth.

Interestingly, it doesn't take much to make people happy, at least in a lab. A small bag of candy or watching a few minutes of a comedy film is often enough. In one case a group of students was shown a short film consisting of bloopers from *The Red Skelton Show* and old television Westerns; this was the "happy" group. Another group was shown a mathematics film, *Area Under a Curve*. This, understandably, was the "not-happy" group. Afterward, both groups were given ten minutes to solve the candle problem. Only 20 percent of those who watched the math film got the candle problem right. But for those who watched the TV bloopers, the success rate was much higher: 75 percent.

Similar effects have been shown in real-world settings, like buying and selling appliances or deciding which car to buy. People in happy moods tend to make their decisions more quickly, with less back-and-forth, and actually enjoy the process more than people who weren't as happy. Even doctors attempting to make a diagnosis fall under the same influence. In one test, a group of doctors was given a small bag of wrapped candy containing Hershey's chocolates; another group got nothing. Both were then asked to look at a patient's history and give a diagnosis of his condition (which was chronic active hepatitis). The doctors who got the candy were quicker than the others to note that the liver was the likely source of trouble.

### Beating a Dead Horse

One thing does not seem to eliminate mistakes, at least not as well as is often assumed. That thing is money. Quite a few studies in recent years have analyzed the way financial incentives affect human behavior, at least the kind of human behavior that can be observed in a lab. The most common result from these experiments is that incentives do not affect average performance. There are, of course, exceptions. On mundane tasks, like filing or other clerical work, money does make a difference. Money will also induce people to endure pain. In one experiment, for instance, students were asked to keep their hands submerged in a tank of ice water for as long as possible. Those who got paid kept their hands submerged nearly three times longer, on average, than those who weren't.

> The most common result is that financial incentives do not affect average performance.

But most of us aren't interested in enduring more pain—or more mundane tasks. Neither are economists. On the kinds of sophisticated tasks that economists are most interested in, like trading in markets or choosing among gambles, the overwhelming finding is that increased incentives do not change average behavior substantively. Generally, what incentives do is prolong deliberation or attention to a problem. People who are offered them will work *harder* on a given problem (in one experiment, their pupils actually grew larger), though they will not necessarily work any *smarter*. Typically, they will pound away at whatever the prevailing strategy is rather than coming up with a new one—effectively, beating a dead horse.

Indeed, in a few tasks, it's been shown that incentives can actually hurt. All of these, interestingly, are tasks that involve judgment or decision. One way this happens is that people become self-conscious about an activity that should be automatic. A good example is choking in sports. Professional basketball players, for instance, sink significantly fewer free-throw shots in play-off games than during the regular season.

### The Currency of Life

After more than a decade of studying what makes people happy, David Schkade told me, he and his colleagues have come to one conclusion: the currency of life isn't money, it's time. When people make major life changes, like moving to a new city or retiring from work, one of their biggest mistakes is not changing the way they use their time. When he lived in Texas, Schkade said, he saw a bumper sticker that summed up this philosophy perfectly: "If You ♥ New York, Take I-30 East."

In other words, if you move to Texas, learn to enjoy the things Texas has to offer. Don't move there expecting to find a great bagel, as you would in New York, or great beaches, as you would in L.A.

Learn to love the rodeo or the Dallas Cowboys or the vast open spaces of west Texas—or else you're likely to be miserable.

Like many of the findings in this book, this one may strike you as obvious. It did me, and I mentioned this to Schkade.

"It is common sense," he agreed. "But people don't do it."

It takes determination and discipline to re-craft a life—which is why, he said, so many retired people end up going back to work. The mistake they make is that they spend their time doing the same old things they've always done and not the new things they thought they were going to do. In his own case, Schkade said, after he moved from Texas to California a few years ago, he made sure to do more of the things that a life in Southern California affords. He golfs more now than he used to, for instance. He's also built a deck onto his house so he can see the sun set over the ocean. And every Sunday morning he and his wife go for a walk on the beach. But in the end, he reminded me, it's not where you live that makes you happy; it's how you use your time.

Forgetting that may be the biggest mistake of all.

# Acknowledgments

This book would not have been possible without the research of the many scholars whose works fill these pages. To them, I am deeply indebted. In particular, I'd like to thank those people who shared with me not only their work, but their time: Laura Beckwith, Marianne Bertrand, Alan Brown, Mike Conlin, Ed Cornell, Karen Daniel, Anders Ericsson, Tom Gilovich, Paul Green, Justin Kruger, Ellen Langer, George Loewenstein, Michael McCloskey, Vicki McCracken, Dan Montello, Dick Neisser, Lynne Reder, Craig Roberts, David Schkade, Paul Schoemaker, Dan Simons, Alex Todorov, Barbara Tversky, Elke Weber, and Jeremy Wolfe.

I am especially grateful to those people who opened their lives and their homes to me. In particular I want to thank Jill Byrne, Norman Einstein, Claire Hewitt, Pete and Jen Marino, and, for a tale I will never forget, Tom Vander Molen.

I also owe a great deal of gratitude to my agents, Jane Dystel and Miriam Goderich. They not only worked diligently to bring this book to life, but offered much-appreciated guidance along the way. Likewise, a deep bow goes to my editor at Broadway Books, Kris Puopolo. Her deft touch and good counsel have improved this book in ways both invisible and immeasurable—as has the tireless work of Stephanie Bowen at Broadway. For their comments on various portions of the book, I am once again in the debt of some old friends: Greg Berg, Kevan Miller, and, most of all, Dr. Jim Lloyd.

Finally, the most obvious acknowledgment of all: any book about mistakes is bound to contain some. For those, the credit belongs entirely to the author.

# References

## Introduction: Why *Do* We Make Mistakes? Because . . .

The story of the Welshmen of St. Brides is taken from Byrne (2000). For the tendencies of right-handed people, see Scharine and McBeath (2002). For the "blue-seven" phenomenon, see Simon (1971) and Simon and Primavera (1972), as well as Kubovy (1977) and Kubovy and Psotka (1976). For more on the power of first impressions, see Kruger, Wirtz, and Miller (2005) and Mathur and Kruger (2007). For the effect of expectations on the weights of truck drivers and dancers, see Christiansen et al. (1983); for North Dakota wine, see Hill (2006); for farmers and global warming, see Weber (1997). For more on forgetting passwords and PINs, see Brown et al. (2004). For dashboard distraction, see Kiley and Eldridge (2002), Vlasic (2008), and Wilson (2008).

A summary of the steps taken by anesthesiologists can be found in Hallinan (2005). For the effectiveness of multi-unit pricing, see Wansink, Kent, and Hoch (1998). The influences used in buying condoms are discussed in Wilson and Brekke (1994). For more on the interesting effects of sleep deprivation, see Belenky (1994a and b) and Williamson and Feyer (2000). For memory underwater, see Godden and Baddeley (1975). For mood and memory, see Bower (1981). And for taking children for a walk in the park, see Wilkinson (1988).

## Chapter 1: We Look but Don't Always See

Helping to prove that people really do buy *Playboy* for the articles, you can find the full Burt Reynolds interview in Linderman (1979). For differing eyewitness accounts by men and women, see Powers, Andriks, and Loftus

(1979) and Loftus et al. (1987). The perceptions of right-handed people are recounted in Scharine and McBeath (2002) and Martin and Jones (1999). Golfers and the differences in their putting gazes are detailed in Vickers (1992 and 1996); the information on Lasik surgery is from Newport (2007); and the quiet-eye period is explained in Janelle and Hillman (2003). The fascinating door experiments demonstrating change blindness can be found at Simons and Levin (1998) and Levin and Simons (1997). Simons's comments are from an author interview. See also Rensink, O'Regan, and Clark (1997). Hewitt's comments are from an author interview.

On the largely automatic processing of information, see Reason (1990), p. 34. For "Turning the Tables" and other illusions, see Shepard (1990) and Shepard and Cooper (1982). Background on Shepard and his experiments can be found at www.nsf.gov/news/frontiers.archive/6-96/6illusio.jsp. Wolfe's comments are from an author interview. For a broader look at the "beer in the refrigerator" problem, see Begley (2005) and Goldberg (2005). The accounts of formerly blind people are in Senden (1960); more recent findings on the ability of formerly blind people to see can be found in Held et al. (2008). For more on not noticing rarely seen items, like guns in airline baggage, see Wolfe, Horowitz, and Kenner (2005); see also Boot et al. (2006). Mammogram rates are from Wolfe, Horowitz, and Kenner (2005); error rates for radiologists are discussed by Berlin (2000), Lorentz et al. (1999), and Muhm et al. (1983). The miss rates at L.A. and O'Hare are reported in Frank (2007). For seized guns and other data, see the Department of Homeland Security Web site at www.dhs.gov/xnews/releases/press_release_0578.shtm.

## Chapter 2: We All Search for Meaning

The research on memory for yearbook photos is Bahrick, Bahrick, and Wittlinger (1975). The well-known penny experiment is from Nickerson and Adams (1979); the British version is Jones (1990). The experiment on Ann Collins and Bristol is from Cohen and Faulkner (1986) and is recounted in Young (1993); also see Cohen (1990). I am indebted to Ann Killion for her recounting of her interview with Joe Theismann. For more on Theismann and his recall of the Einsteins, see Wulf (1992). For a review of tip-of-the-tongue experiences, see Brown (1991). For the Liza

Minnelli example, see Yarmey (1973); for more on proper nouns and getting names unstuck, see Brennen et al. (1990). Some of Norman Einstein's comments are from an author interview.

The feats behind the famous forgetting curve are detailed in Ebbinghaus (1964); they are also outlined in Mook (2004). The story of the long-distance runner with the amazing memory is documented by Chase and Ericsson (1981). The poll of three thousand people is mentioned in Reuters (2007b). Tom Vander Molen's story is from an author interview. The survey of four hundred adults who had recently found an object they had lost is in Tenney (1984). For unusualness making a hiding place more forgettable, see Winograd and Soloway (1986). The *New York Times* example is mentioned by Campanelli (2006), and the forgetting rate for passwords is from Brown et al. (2004). See also Brown and Rahal (1994). For calls to help desks, see Fielding (2003).

For greater depth of processing and traits, see Bower and Karlin (1974); for trait recognition, see Winograd (1976); and for the importance of hair and other factors in recognition, see Shepherd, Davies, and Ellis (1981). For newborns' preferences for faces, see Johnson et al. (1991). An interesting paper on quickly extracting trait information from faces is Willis and Todorov (2006). June Siler's account is based on court records, media accounts, and videotaped statements Siler made at a seminar at Northwestern University Law School. See in particular Possley (2006). For a summary of the DNA evidence study, see *Harper's* (2007). For more on remembering pretty faces, see Cross, Cross, and Daly (1971); and for the ugly faces of crime—and criminals—see Mocan and Tekin (2006) and Morin (2006).

## Chapter 3: We Connect the Dots

For more on inferences of competence based solely on facial appearance, see Todorov et al. (2005) and Willis and Todorov (2006). The West Point study is Mueller and Mazur (1996). Subtle changes in facial attractiveness during menstrual periods are described by Roberts et al. (2004). The study of lap dancers is from Miller, Tybur, and Jordan (2007); the effect of fragrances is from Spangenberg et al. (2006). The study of expensive wine was done by Plassman, O'Doherty, Shiv, and Rangel (2008); the research on the $2.50 placebo is from Waber, Shiv, Carman, and Ariely (2008). The relationship between black uniforms and penalties is outlined in Frank and Gilovich (1988). The merits of changing one's answers on a test have been studied for as long as most of us have been alive. An informative paper—and a recent one—is Kruger, Wirtz, and Miller (2005). For a review, see Prinsell, Ramsey, and Ramsey (1994). For the Monty Hall problem, see Gilovich, Medvec, and Chen (1995). Kruger's comments are from an author interview.

## Chapter 4: We Wear Rose-Colored Glasses

The story of Steve Wynn and Picasso's mistress can be found in many places. In addition to Ephron's account, see Zambito (2007), Clarke (2006), and Paumgarten (2006). The Ohio Wesleyan study is Bahrick, Hall, and Berger (1996), and is recounted in Neisser and Hyman (2000). For parents and their memories of their parenting methods, see Robbins (1963). For recognizing enhanced versions of our own faces, see Angier (2008) and Epley and Whitchurch (2008). The Kahneman quotation is from Schrage (2003). For a fuller treatment of John Dean's testimony, see Neisser (1981); some of Neisser's comments are from an author interview, as are Dean's.

The quotation from Wohlstetter (1962) is from p. 387. The seminal papers on hindsight research are Fischhoff and Beyth (1975) and Fischhoff (1975). For an overview of developments since then, see Fischhoff (2007). For memory of lifetime sex partners, see Brown and Sinclair (1999), and for the most recent U.S. data see Fryar et al. (2007). Gamblers' memories of their bets are detailed in Gilovich (1983) and in an author interview.

For voters and their impressions, see Ballew and Todorov (2007) and Todorov et al. (2005). For an amusing take on bias in other people (but not ourselves), see Gilbert (2006). Details of prescription drug use by Americans are taken from U.S. Department of Health and Human Services (2004). Data on the cost of new drugs are taken from National Institute for Health Care Management (2002). For pharmaceutical company spending per doctor, see Gibbons et al. (1998). The experiment on the ineffectiveness of disclosing conflicts of interest is in Cain, Loewenstein, and Moore (2005); see also Dunleavey (2007). For more on moral licensing, see Monin and Miller (2001). For Jesse Jackson's remark, see Associated Press (2008b). Loewenstein's comments are from an author interview.

## Chapter 5: We Can Walk and Chew Gum—but Not Much Else

The crash of Flight 401 is documented in National Transportation Safety Board (1973). For more on Controlled Flight into Terrain (CFIT), see Shappell and Wiegmann (2001 and 2003) and *Air Safety Week* (2005); Origin of term is from Bateman (2008) and Carbaugh (2008). For stats on CFIT, see Matthews (1997). For the experience of the Air Force with CFIT, see Moroze and Snow (1999). For the constant interruptions faced by office workers, see Gonzalez and Mark (2004). For the difficulties entailed in multitasking, see Klein (2007) and Pashler (1994) as well as Jiang (2004) and Jiang, Saxe, and Kanwisher (2004). For the hidden costs of switching from task to task, see Einstein et al. (2003); taking fifteen minutes to regain a deep state of concentration comes from Douglas et al. (2005). The detail about Microsoft's workers is recounted in Lohr (2007). For e-mailing distractions generally, see Searcey (2008).

The Army study is Middlebrooks, Knapp, and Tillman (1999). The classic work on inattentional blindness is Mack and Rock (1998). The details of the bus crash are in National Transportation Safety Board (2006a), in documents obtained under a Freedom of Information Act request by the author, and in Gowen and Arzua (2004). For driver distraction and other findings from NHTSA, see Dingus et al. (2006) and Klauer et al. (2006). Klauer's comments are from an author interview. For entry times on navigation devices, see Tsimhoni, Smith, and Green (2004). For driver inattention and its role in crashes, see Wang, Knipling, and Goodman (1996). The fatal crash in upstate New York is detailed in Vlasic (2008), and New York City's proposed ban is mentioned in Konigsberg (2008). The use of night-vision equipment in cars is from Welsh (2008); for dashboard distraction generally, see Kiley and Eldridge (2002), Wilson (2008), and Vlasic (2008). For Nardelli's comments, see Maynard (2007). For more on Sync, including comments by Gates and Fields, see www.syncmyride.com. For more details on the effects that the use of hands-free devices while driving can have on the brain, see Harbluk, Noy, and Eizenman (2002).

For more on the limits of multitasking in general and the Stroop effect in particular, see Manhart (2004). For interruptions and BLIS, see Mateja (2007). For growth of older drivers, see U.S. Government Accountability Office (2007). For shrinkage of visual field with age, see Wolf (1967). For details on task recovery time and the perils of interrupting someone in the middle of a task, see Monk, Boehm-Davis, and Trafton (2004). For driver workload, see Recarte and Nunes (2000 and 2003). Details on Volvo's Intelligent Driver Information System and similar devices can be found in Green (2004) and in Peirce and Lappin (2006). The number of trucks is taken from the *Statistical Abstract of the United States, 2006.* The account of the accident involving Linda Camacho is taken from court documents and from an extensive article by Jones, Becka, LaFleur, and McGonigle (2006). Details regarding the Qualcomm e-mail system were taken from company documents on file with the Securities and Exchange Commission.

## Chapter 6: We're in the Wrong Frame of Mind
For coverage of the Van Iveren incident, see Doege (2007) and Doege and Rinard (2007). The framing effect of music on wine purchases is succinctly outlined by North, Hargreaves, and McKendrick (1997). A good overview of the research by Kahneman and Tversky can be found in Mook (2004). Slovic's comments are from Kluger (2006). The study of fourth-down decision making is Romer (2006).

For framing effects while shopping, see Mindlin (2007) and Morewedge, Holtzman, and Epley (2007). The case of EntreMed was detailed by Huberman and Regev (2001); the company's price spike was reported by Fisher (1998); the article in *Nature* is Boehm et al. (1997). The relationship between childbirth and anesthesia can be found in Christensen-Szalanski (1984). The cancellations at Jenny Craig are reported in Barnes and Petersen (2001). For an interesting take on picking lowbrow movies and junk food, see Read, Loewenstein, and Kalyanaraman (1999) and Read and van Leeuwen (1998). For a broader context on the way time affects our decisions, see Ferraro et al. (2005), and for time and dieting see Herman and Polivy (2003). The research on catalog orders for cold-weather gear can be found at Conlin, O'Donoghue, and Vogelsang (2006). The information on the South African loan experiment featuring pictures of the opposite sex is taken from Bertrand, Karlan, Mullainathan, Shafir, and Zinman (2006); Mullainathan's comments are from Lambert (2006).

For more on the power of anchoring, see Strack and Mussweiler (1997). Much has been written about prices in grocery stores; for an article on the more interesting effects of framing, see DelVecchio, Krishnan, and Smith (2007). For using information in the form in which it is displayed, see Payne, Bettman, and Johnson (1993). For the order of ballots, see Ho and Imai (2006). For the value of a first offer in negotiation, see Galinsky and Mussweiler (2001). For multiple-unit pricing, see Wansink, Kent, and Hoch (1998). Given the recent troubles in the real estate market, the work of Northcraft and Neale (1987) makes even more interesting reading now than it did twenty years ago. For the effect of "sales" items in grocery stores, see Jargon, Zimmerman, and Kesmodel (2008). For overcoming anchoring effects, see Mussweiler, Strack, and Pfeiffer (2000).

## Chapter 7: We Skim

For the unicorn story, see Raghavan (2004); the correction appeared a week later and is appended to the original story. For missing the letter *e*, see Corcoran (1967). For investors nodding off on Fridays, see DellaVigna and Pollet (2008). For musical sight-reading, see Sloboda (1988). The story of Boris Goldovsky is taken from Wolf (1976). The NASA correction is detailed in Agence France Press (2008); the error at the Smithsonian is from Associated Press (2008a); and for the story of the Russian sub, see Parfitt (2007). The Halloween hanging was reported by Merriweather (2005) and Associated Press (2005). The explanation for using a washing machine is from Bransford and Johnson (1972). The research on taking children for a walk in the park is from Wilkinson (1988); underwater memory is from Godden and Baddeley (1975); and the memory for happy characters is from Bower (1981).

## Chapter 8: We Like Things Tidy

For map distortions, see Tversky's work, especially her 1981 paper. Milgram's research is detailed in Milgram (1974) and Milgram and Jodelet (1976). For navigation of bees, see Srinivasan et al. (1996) and Esch and Burns (1996). For a larger context on pruning inconvenient details from our memories, see Fischhoff and Beyth (1975). "The War of the Ghosts" is from Bartlett (1964). For more on what makes a price memorable, see Mindlin (2006) and Vanhuele, Laurent, and Dreze (2006). For details on recalling the words of "The Star-Spangled Banner" and the role of music in general, see Rubin (1977 and 1994) and Wallace and Rubin (1998a and b). For details on the difficulty of verbatim reproduction, see Wade and Clark (1993). For untruths in conversation, see DePaulo et al. (1996), Tversky and Marsh (2000), Feldman, Forrest, and Happ (2002), Dudoko-vich, Marsh, and Tversky (2004), and Tversky (2004).

## Chapter 9: Men Shoot First

For Finnish drivers and their stock portfolios, see Grinblatt and Keloharju (2008). For investors who trade the most and earn the least, see Barber and Odean (2000). For trading results of men and women, see Barber and Odean (2001). For men and women estimating their own IQs, see Reilly and Mulhern (1995). For overestimating their own attractiveness, see Gabriel, Critelli, and Ee (1994). For girls underestimating math grades, see Beyer (1998). For more on men, women, and starting wars, see Johnson et al. (2006). For Tenet's slam-dunk quotation, see Leibovich (2004).

The study of soldiers' ability to tell friend from foe is by Johnson and Merullo (1999). Several studies have examined the issue of women being more risk averse than men; one of the more recent, Harris, Jenkins, and Glaser (2006), mentions higher numbers of deaths for men from events like drowning and accidental poisonings. For a fuller account of Elke Weber's research, see Weber, Blais, and Betz (2002). For interesting studies on the lies men and women tell, see Feldman, Forrest, and Happ (2002) and DePaulo et al. (1996). For an overview of differences between college men and women, see Maccoby and Jacklin (1974), p. 154. For the lottery ticket experiment, see Langer (1975).

For women and computing, see Beckwith (2007a) and Associated Press (2007b). Navigational failures and the differences noted between male and female drivers are detailed in King (1986). The wayfinding abilities of boys and girls are laid out in the books and articles of Ed Cornell and colleagues, which are listed in the bibliography. They discuss the fascinating concept of children's "home range," as does Matthews (1987 and 1992). For the performance of men and women on wayfinding tasks, see Cornell, Sorenson, and Mio (2003). On the subjects of men and women and their sense of direction and "environmental confusion" and various wayfinding strategies, see the papers of Carol Lawton. Also see Choi, McKillop, Ward, and L'Hirondelle (2006) and Dabbs et al. (1998). The comments of Beckwith and Montello are from author interviews.

## Chapter 10: We All Think We're Above Average

For the Princeton study, see Gilbert (2006). DellaVigna's comments are from an author interview. For the pro shop putting example, see Burson (2007) and Wu (2007). Details of NutriSystem's performance are from Hallinan (2007). Information on the company's customers can be found in the company's 10-K, in its Investor Presentation of March 27, 2007, and in other filings with the Securities and Exchange Commission. On paying not to go to the gym, see DellaVigna and Malmendier (2005). Overconfidence and credit card teaser rates are detailed in Ausubel (1999) and DellaVigna (2007b).

Marksmanship in the U.S. Army was documented by Schendel, Morey, Granier, and Hall (1983). The poor calibration of University of Wisconsin students is taken from Glenberg and Epstein (1985). A fair amount has been written on how well-calibrated weather forecasters tend to be; one of the more thorough treatments is Murphy and Winkler (1984). For the history of using probabilistic forecasting, see Hallenbeck (1920). For more on the importance of feedback, see Norman (1988) and Biederman (1987). For the response of men and women to negative feedback, see Beyer (1998), Roberts (1991), and Roberts and Nolen-Hoeksema (1989). Information on Dexter Shoe can be found in Warren Buffett's 2007 and 1993 letters to the shareholders of Berkshire Hathaway Inc., which are archived on the company's Web site.

For the illusion of control, see Langer (1975). For more on the beguiling power of information, see Reder and Anderson (1980) for insights on the effectiveness of summaries versus full texts. The section on Jill Byrne is from an author interview at the racetrack. Slovic's 1973 study on horse handicappers is a classic and is available at his home page: www.decisionresearch.org/people/slovic/. The boast by D. R. Horton's chief executive is in Hagerty and Dunham (2005); the information about the company's loss is from Corkery (2007). Finally, Russo and Schoemaker have written an insightful book about managerial decision making, *Decision Traps* (1990). This was followed by a jointly written paper, on managing overconfidence, in 1992.

## Chapter 11: We'd Rather Wing It

Details of the PGA study on the perils of putting are taken from Diaz (1989). For clinical psychologists and their secretaries, see Menand (2005). For more on the miserable performance of securities analysts and other overconfident types, see DellaVigna (2007b) and Dremen and Berry (1995). Colin Camerer's assessment is taken from Camerer and Johnson (1991); for more on the subject, see Bishop and Trout (2002), Dawes, Faust, and Meehl (1989), and Goldberg (1968). For predictions by political experts (and their lack of modesty), see Tetlock (1998).

Ericsson's comments are taken in part from an author interview; over the years his writings on expertise have become classics on the subject; see especially Ericsson and Smith (1991). For more on the role played by IQ and innate characteristics, see Starkes and Ericsson (2003) and especially Ericsson, Roring, and Nandagopal (2007). This last article is an excellent overview on expertise in sports and also provides interesting information on pattern recognition. For details on the memories of chess masters, see Chase and Simon (1973). See also Charness (1991). For fixing notational errors on the fly, see Ericsson and Smith (1991), pp. 31 and 156.

An overview of Tolman's work can be found in Mook (2004), but Tolman's 1948 paper makes for great reading, even sixty years later. McConnell's experience in the seventh grade is recounted in his 1996 book, *Rapid Development*. The electrical plug example is taken from Hull, Wilkins, and Baddeley (1988). The recall of mock jurors was reported in Hastie, Schkade, and Payne (1999). The nail-gun study can be found at *Morbidity and Mortality Weekly Report* (2007). For an eye-opening look at similar self-inflicted injuries involving medications taken at home, see Associated Press (2008c). For the number of "readily discriminable" objects, see Norman (1988). The role of Cissell and the advent of clothing labels is detailed in Akst (2001). The length of the Mercedes owner's manual is detailed in Sabatini (2006). The Subaru example is from Mayer (2002). Error rates for car seat installation can be found at National Highway Traffic Safety Administration (1986, 2004, and 2006); Decina's comments are from an author interview.

Much has been written about the straight-down effect; for original research, see McCloskey, Washburn, and Felch (1983). For an interesting read about people's intuitive understanding of bodies in motion and pre-Newtonian physics, see McCloskey's 1983 article in *Scientific American*. The water experiments are well explained in Mook (2004). If you want the originals, go to Luchins and Luchins (1950) and Luchins (1942). Details of the well-known candle-box-and-tack trick can be found at Duncker (1939).

## Chapter 12: We Don't Constrain Ourselves

For more on constraints and affordances, see Norman (1988) and Gibson (1979). For the harrowing story of the Quaid twins, see the excellent coverage of the *Los Angeles Times* at Ornstein (2007 and 2008). Details regarding EDTA can be found at Dooren (2008). The section on aviation fixes draws heavily from McCartney (2006). For more on BMW's iDrive system, see Meiners (2004) and Cobb (2002). For the improvements made to heparin vials, as well as the changes implemented by Methodist Hospital, see Landro (2008). For surgeons marking surgical sites, see Davis (2006). The bartender study can be found at Beach (1988).

The study of naval aviators and doctors is Gaba et al. (2003). For more on the "golden age" of aviation safety, see Wald (2007). Misdiagnosis and autopsies are discussed in Leonhardt (2006), Brownlee (2007), Shojania et al. (2003), and Lundberg (1998). The survey of pilots and surgeons regarding the questioning of senior staff members is from Sexton, Thomas, and Helmreich (2000). For more on the development of Crew Resource Management and the crash in Portland, see McCartney (2005) and National Transportation Safety Board (1979). For application of CRM principles in a corporate setting, see Pearce (2008). Charles Vincent's case study can be found at Vincent (2003).

**References**

**Chapter 13: The Grass _Does_ Look Greener**

Some of Schkade's comments are from an author interview. Schkade and Kahneman's 1998 study was followed by two other relevant papers, Kahneman et al. (2004 and 2006). Loewenstein and Schkade also teamed up in a 2003 paper, "Wouldn't It Be Nice? Predicting Future Feelings." It includes examples involving smokers and cancer patients. For more on rebates, see Grow (2005). For more on unused gift cards, see Merrick (2008), Deloitte (2007), Clothier (2006), and TowerGroup (2006); and for an interesting twist, see Thurm and Tam (2008). For spousal predictions, see Davis, Hoch, and Ragsdale (1986). Net-a-Porter.com is from Tan (2007). Our preference for changeable outcomes is outlined by Gilbert and his colleagues, whose papers (which are often very funny and well worth reading) are noted in the bibliography. For details of the colostomy study, see Smith, Loewenstein, Jankovich, and Ubel (2007). For happiness levels of the handicapped, see Diener and Diener (1996). For a consumer point of view on rebates, see Barlyn (2007); and for a telling example of their power, see Miller (2006).

**Conclusion**

For more on the sloppy handwriting of doctors, see Caplan (2007). For studies on children and adults not tolerating mistakes, see Tugend (2007); see also, generally, Dweck (1999). The Royal Dutch Shell example is in Russo and Schoemaker (1992). The power of negative thinking is Gawande (2007). The banning of iPods and similar devices for runners in races is taken from Macur (2007). The CIA study is from Heuer (1999).

The effects of sleep deprivation are discussed in Harrison and Horne (2000), Williamson and Feyer (2000), Pilcher and Huffcutt (1996), Belenky (1994a and b), and Haslam and Abraham (1987). For sleep-eating, see Beck (2008). For pilots falling asleep mid-flight, see Levin and Heath (2007). For JetBlue's experiment, see Pasztor and Carey (2006). The efforts of Continental and American airlines are detailed in Pasztor (2007). For sleepy doctors, see Fackelmann (2006) and Barger et al. (2006).

The role of happiness is discussed in the papers of Isen and her colleagues. For the connection between happiness and creativity, see also Greene and Noice (1988). For an excellent review of experiments on the effect that financial incentives can have on human behavior, see Camerer and Hogarth (1999). See also Read (2005) and Hertwig, Pachur, and Kurzenhauser (2005). For more on incentives and working harder versus working smarter, see Payne, Bettman, and Johnson (1993), especially pp. 111 and 156–57. For enduring pain in ice water, see Baker and Kirsch (1991). The pupil-dilation experiment is Kahneman and Peavler (1969). Free-throw shooting by professional basketball players is discussed in Camerer (1998). Schkade's comments are from an author interview.

# Bibliography

Adams, Richard J., and K. Anders Ericsson. 1992. *Introduction to Cognitive Processes of Expert Pilots.* U.S. Department of Transportation, Federal Aviation Administration, Report No. DOT/FAA/RD-92/12.

Agence France Press. 2008. German Schoolboy, 13, Corrects NASA's Asteroid Figures: Paper. April 15.

Ahlers, Mike M. 2007. Air Controller Fatigue Contributed to 4 Mishaps. CNN.com, April 11.

*Air Safety Week.* 2005. Despite Headway, CFIT Remains Persistent, Deadly Threat. 19 (2), Jan. 10.

Akst, Daniel. 2001. Read the Instructions. *Industry Standard*, June 11.

Alexander, Amy L., Thomas E. Nygren, and Michael A. Vidulich. 2000. *Examining the Relationship Between Mental Workload and Situation Awareness in a Simulated Air Combat Task.* Air Force Research Laboratory, Wright-Patterson Air Force Base, Dayton, Ohio.

Anderson, Jenny, and Vikas Bajaj. 2008. Merrill Tries to Temper the Pollyannas in Its Ranks. *New York Times,* May 15, p. C1.

Angier, Natalie. 2008. Mirrors Don't Lie. Mislead? Oh, Yes. *New York Times,* July 22, p. D1.

Ariely, Dan, and Klaus Wertenbroch. 2002. Procrastination, Deadlines, and Performance: Self-Control by Precommitment. *Psychological Science* 13 (3), pp. 219–24.

Arkes, Hal R. 1991. Costs and Benefits of Judgment Errors: Implications for Debiasing. *Psychological Bulletin* 110 (3), pp. 486–98.

Arkes, Hal R., Robyn M. Dawes, and Caryn Christensen. 1986. Factors Influencing the Use of a Decision Rule in a Probabilistic

Task. *Organizational Behavior and Human Decision Making* 37, pp. 93–110.

Arkes, Hal R., et al. 1981. Hindsight Bias Among Physicians Weighing the Likelihood of a Disease. *Journal of Applied Psychology* 66, pp. 252–54.

Armstrong, David. 2007. Your Doctor's Business Ties Are Your Business, Too. *Wall Street Journal*, Nov. 20, p. D1.

Associated Press. 2005. Body Hanging from Tree Mistaken for Halloween Decoration. Oct. 28.

————. 2007a. FDA Says Pills Can Cause "Sleep-Driving." March 14.

————. 2007b. Subtle Software Changes May Narrow Gender Gap. *Chicago Tribune*, Sept. 24, sec. 3, p. 5.

————. 2008a. 5th-Grader Finds Mistake at Smithsonian. April 3.

————. 2008b. Fox: Jackson Used N-Word in Off-Air Remarks. July 16.

————. 2008c. Fatal Medication Errors at Home Rise Sharply. July 29.

Atchley, P., and J. Dressel. 2004. Conversation Limits the Functional Field of View. *Human Factors* 46 (4), pp. 664–73.

Ausubel, Lawrence M. 1999. Adverse Selection in the Credit Card Market. Working paper, University of Maryland.

Bahrick, Harry P., Phyllis O. Bahrick, and Roy P. Wittlinger. 1975. Fifty Years of Memory for Names and Faces: A Cross-Sectional Approach. *Journal of Experimental Psychology: General* 104 (1), pp. 54–75.

Bahrick, Harry P., Lynda K. Hall, and Stephanie A. Berger. 1996. Accuracy and Distortion in Memory for High School Grades. *Psychological Science* 7, pp. 265–71.

Baker, S. L., and I. Kirsch. 1991. Cognitive Mediators of Pain Perception and Tolerance. *Journal of Personality and Social Psychology* 61, pp. 504–10.

Ballew, Charles C., II, and Alexander Todorov. 2007. Predicting Political Elections from Rapid and Unreflective Face Judgments. *Proceedings of the National Academy of Sciences* 104 (46), pp. 17948–53.

Barber, Brad M., and Terrance Odean. 2000. Trading Is Hazardous to Your Wealth: The Common Stock Investment Performance of Individual Investors. *Journal of Finance* 55 (2), pp. 773–806.

————. 2001. Boys Will Be Boys: Gender, Overconfidence, and Common Stock Investment. *Quarterly Journal of Economics*, Feb., pp. 261–92.

————. 2008. All That Glitters: The Effect of Attention and News on the

Buying Behavior of Individual and Institutional Investors. *Review of Financial Studies* 21, pp. 785–818.

Barger, Laura K., et al. 2006. Impact of Extended-Duration Shifts on Medical Errors, Adverse Events, and Attentional Errors. *Public Library of Science—Medicine,* available online.

Barlyn, Suzanne. 2007. Cranky Consumer: Waiting for Rebate Checks to Arrive. *Wall Street Journal,* May 10, p. D2.

Barnes, Brooks, and Andrea Petersen. 2001. As Priorities Change, Some Question Why They Eschew the Fat. *Wall Street Journal,* Oct. 5, p. 1.

Barras, Colin. 2008. "Sexy" Voice Gives Fertile Women Away. *New Scientist,* May 1, p. 14.

Bartlett, F. C. 1964. *Remembering: A Study in Experimental and Social Psychology.* Cambridge, U.K.: Cambridge University Press. (1st ed., 1932.)

Bateman, Don. 2008. Personal correspondence.

Beach, King D. 1988. The Role of External Mnemonic Symbols in Acquiring an Occupation. In *Practical Aspects of Memory: Current Research and Issues.* Vol. 1: *Memory in Everyday Life,* edited by M. M. Gruneberg, P. E. Morris, and R. N. Sykes. Chichester, U.K.: Wiley.

Beck, Melinda. 2008. To Cut Risks of Sleeping Pills, Hide Car Keys, Unplug Phone. *Wall Street Journal,* May 6, p. D1.

Beckwith, Laura A. 2007a. Gender HCI Issues in End-User Programming. Ph.D. diss., Oregon State University.

———. 2007b. Author interview.

Begley, Sharon. 2005. Security's Blind Spot. *Wall Street Journal,* Dec. 30, p. A11.

Belenky, Gregory, et al. 1994a. Subjective Fatigue of C-141 Aircrews During Operation Desert Storm. *Human Factors* 36 (2), pp. 339–49.

———. 1994b. The Effects of Sleep Deprivation on Performance During Continuous Combat Operations. In *Food Components to Enhance Performance,* edited by B. Marriott Washington, D.C.: National Academy Press.

Berlin, Leonard. 2000. Hindsight Bias. *American Journal of Roentgenology* 175, pp. 597–601.

Bertrand, Marianne. 2007. Author interview.

Bertrand, Marianne, Dean Karlan, Sendhil Mullainathan, Eldar Shafir, and Jonathan Zinman. 2006. Pricing Psychology: A Field Experiment in the Consumer Credit Market. Mimeo, University of Chicago.

Beyer, Sylvia. 1998. Gender Differences in Self-Perception and Negative Recall Biases. *Sex Roles* 38 (1/2), pp. 103–33.

Biederman, Irving. 1987. Recognition-by-Components: A Theory of Human Image Understanding. *Psychological Review* 94, pp. 115–47.

Bishop, Michael A., and J. D. Trout. 2002. 50 Years of Successful Predictive Modeling Should Be Enough: Lessons for Philosophy of Science. *Philosophy of Science* 69, pp. S197–S208.

Blackstone, John. 2008. Bringing Back Bridge. CBS News, Feb. 16, available online at cbsnews.com.

Blass, Thomas. 2004. *The Man Who Shocked the World: The Life and Legacy of Stanley Milgram.* New York: Basic Books.

Boehm, Thomas, et al. 1997. Antiangiogenic Therapy of Experimental Cancer Does Not Induce Acquired Drug Resistance. *Nature* 390, pp. 404–7.

Bonner, Sarah E., S. Mark Young, and Reid Hastie. 1996. Financial Incentives and Performance in Laboratory Tasks: The Effects of Task Type and Incentive Scheme Type. Unpublished manuscript. University of Southern California Department of Accounting.

Boot, Walter R., et al. 2006. Detecting Transient Changes in Dynamic Displays: The More You Look, the Less You See. *Human Factors* 48 (4), pp. 759–73.

Bower, G. H. 1981. Mood and Memory. *American Psychologist* 36, pp. 129–48.

Bower, G. H., and M. H. Karlin. 1974. Depth of Processing Pictures of Faces and Recognition Memory. *Journal of Experimental Psychology* 103, pp. 751–57.

Boynton, Robert D., Brian F. Blake, and Joe N. Uhl. 1983. Retail Price Reporting Effects in Local Food Markets. *American Journal of Agricultural Economics* 65 (1), pp. 20–29.

Bransford, John D., and Marcia K. Johnson. 1972. Contextual Prerequisites for Understanding: Some Investigations of Comprehension and Recall. *Journal of Verbal Learning and Verbal Behavior* 11, pp. 717–26.

Brennen, Tim, et al. 1990. Resolving Semantically Induced Tip-of-the-Tongue States for Proper Nouns. *Memory & Cognition* 18 (4), pp. 339–47.

Brouwer, W., et al. 1991. Divided Attention in Experienced Young and

Older Drivers: Lane Tracking and Visual Analysis in a Dynamic Driving Simulator. *Human Factors* 33 (5), pp. 573–82.

Brown, Alan S. 1991. A Review of the Tip-of-the-Tongue Experience. *Psychological Bulletin* 109 (2), pp. 204–23.

————. 2007. Author interview.

Brown, Alan S., et al. 2004. Generating and Remembering Passwords. *Applied Cognitive Psychology* 18, pp. 641–51.

Brown, Alan S., and Tamara A. Rahal. 1994. Hiding Valuables: A Questionnaire Study of Mnemonically Risky Behavior. *Applied Cognitive Psychology* 8, pp. 141–54.

Brown, Norman R., and Robert C. Sinclair. 1999. Estimating Number of Lifetime Sexual Partners: Men and Women Do It Differently. *Journal of Sex Research* 36 (3), pp. 292–97.

Brownlee, Shannon. 2007. *Overtreated: Why Too Much Medicine Is Making Us Sicker and Poorer.* New York: Bloomsbury USA.

Bruce, Vicki, and Andy Young. 1998. *In the Eye of the Beholder: The Science of Face Perception.* Oxford: Oxford University Press.

Burson, Katherine A. 2007. Consumer-Product Skill Matching: The Effects of Difficulty on Relative Self-Assessment and Choice. *Journal of Consumer Research* 34, pp. 104–10.

Byrne, Caroline. 2000. Confused Vigilantes Attack Doctor's Home. *Chicago Sun-Times,* Aug. 31, p. 34.

Byrne, Jill. 2007. Author interview.

Cain, Daylian M., George Loewenstein, and Don A. Moore. 2005. The Dirt on Coming Clean: Perverse Effects of Disclosing Conflicts of Interest. *Journal of Legal Studies* 34, pp. 1–25.

Camerer, Colin. 1998. Behavioral Economics and Nonrational Decision Making in Organizations. In *Decision Making in Organizations,* edited by J. Halpern and B. Sutton. Ithaca, N.Y.: Cornell University Press.

Camerer, Colin, Linda Babcock, George Loewenstein, and Richard Thaler. 1997. Labor Supply of New York City Cabdrivers: One Day at a Time. *Quarterly Journal of Economics* 112, pp. 407–41.

Camerer, Colin, and Robin Hogarth. 1999. The Effects of Financial Incentives in Experiments: A Review and Capital-Labor-Production Framework. *Journal of Risk and Uncertainty* 19, pp. 7–42.

Camerer, Colin, and Eric J. Johnson. 1991. The Process-Performance

Paradox in Expert Judgment: How Can Experts Know So Much and Predict So Badly? In *Toward a General Theory of Expertise: Prospects and Limits*, edited by K. Anders Ericsson and Jacqui Smith. Cambridge, U.K.: Cambridge University Press.

Camerer, Colin, George Loewenstein, and Martin Weber. 1989. The Curse of Knowledge in Economic Settings: An Experimental Analysis. *Journal of Political Economy* 97 (5), pp. 1232–54.

Camerer, Colin, and Dan Lovallo. 1999. Overconfidence and Excess Entry: An Experimental Approach. *American Economic Review* 89 (1), pp. 306–18.

Campanelli, John. 2006. I Forget My Password! With More and More Codes to Clutter Our Brains, It's a Wonder We Don't All Crash. *Plain Dealer,* Sept. 3, p. L1.

Caplan, Jeremy. 2007. Cause of Death: Sloppy Doctors. *Time,* Jan. 15.

Carbaugh, David C. 2008. Personal correspondence.

Cash, James. 2005. Specialist's Factual Report of Investigation. National Transportation Safety Board, Accident No. IAD05FA023, Oct. 18.

Chapman, Gretchen B., and Brian H. Bornstein. 1996. The More You Ask for, the More You Get: Anchoring in Personal Injury Verdicts. *Applied Cognitive Psychology* 10, pp. 519–40.

Chapman, Loren J., and Jean P. Chapman. 1967. Genesis of Popular but Erroneous Psychodiagnostic Observations. *Journal of Abnormal Psychology* 72 (3), pp. 193–204.

Charness, Neil. 1991. Expertise in Chess: The Balance Between Knowledge and Search. In *Toward a General Theory of Expertise: Prospects and Limits,* edited by K. Anders Ericsson and Jacqui Smith. Cambridge, U.K.: Cambridge University Press.

Chase, William G., and K. Anders Ericsson. 1981. Skilled Memory. In *Cognitive Skills and Their Acquisition,* edited by John R. Anderson. Hillsdale, N.J.: Lawrence Erlbaum Associates.

Chase, William G., and Herbert A. Simon. 1973. Perception in Chess. *Cognitive Psychology* 4, pp. 55–81.

Choi, Jean, Erin McKillop, Michael Ward, and Natasha L'Hirondelle. 2006. Sex-Specific Relationships Between Route-Learning Strategies and Abilities in a Large-Scale Environment. *Environment and Behavior* 38 (6), pp. 791–801.

Christensen-Szalanski, J. J. J. 1984. Discount Functions and the Measurement of Patient Values: Women's Decisions During Childbirth. *Medical Decision-Making* 4, pp. 41–48.

Christensen-Szalanski, J. J. J., and James B. Bushyhead. 1981. Physicians' Use of Probabilistic Information in a Real Clinical Setting. *Journal of Experimental Psychology: Human Perception and Performance* 7 (4), pp. 928–35.

Christiansen, R. E., et al. 1983. Influencing Eyewitness Descriptions. *Law and Human Behavior* 7, pp. 59–65.

Clarke, Norm. 2006. Wynn Accidentally Damages Picasso. *Las Vegas Review-Journal,* Oct. 17, p. 3A.

Clothier, Mark. 2006. Retailers Find Profit Windfall—Unused Gift Cards. Bloomberg News, Feb. 27.

CNN Interactive. 1997. Lack of Sleep America's Top Health Problem, Doctors Say. March 17.

Cobb, James G. 2002. Menus Behaving Badly. *New York Times,* May 12.

Cohen, Gillian. 1990. Why Is It Difficult to Put Names to Faces? *British Journal of Psychology* 81, pp. 287–97.

Cohen, Gillian, and Dorothy Faulkner. 1986. Memory for Proper Names: Age Differences in Retrieval. *British Journal of Developmental Psychology* 4, pp. 187–97.

Conlin, Michael, Ted O'Donoghue, and Timothy Vogelsang. 2006. Projection Bias in Catalog Orders. *American Economic Review* 97 (4), pp. 1217–49.

Corcoran, D. W. J. 1967. Acoustic Factor in Proof Reading. *Nature* 214, pp. 851–52.

Corkery, Michael. 2007. Home Myths Meet Reality. *Wall Street Journal,* Aug. 4–5, p. B1.

Cornell, Edward H. 2007. Author interview.

Cornell, Edward H., and Deborah H. Hay. 1984. Children's Acquisition of a Route via Different Media. *Environment and Behavior* 16 (5), pp. 627–41.

Cornell, Edward H., C. Donald Heth, and Lorri S. Broda. 1989. Children's Wayfinding: Response to Instructions to Use Environmental Landmarks. *Developmental Psychology* 25 (5), pp. 755–64.

Cornell, Edward H., C. Donald Heth, and Wanda L. Rowat. 1992.

Wayfinding by Children and Adults: Responses to Instructions to Use Look-Back and Retrace Strategies. *Developmental Psychology* 28 (2), pp. 328–36.

Cornell, Edward H., Autumn Sorenson, and Teresa Mio. 2003. Human Sense of Direction and Wayfinding. *Annals of the Association of American Geographers* 93 (2), pp. 399–425.

Cross, John F., Jane Cross, and James Daly. 1971. Sex, Race, Age, and Beauty as Factors in Recognition of Faces. *Perception & Psychophysics* 10 (6), pp. 393–96.

Dabbs, James M., Jr., et al. 1998. Spatial Ability, Navigation Strategy, and Geographic Knowledge Among Men and Women. *Evolution and Human Behavior* 19, pp. 89–98.

Dana, Jason, and George Loewenstein. 2003. A Social Science Perspective on Gifts to Physicians from Industry. *Journal of the American Medical Association* 290 (2), pp. 252–55.

Davis, Harry L., Stephen J. Hoch, and E. K. Easton Ragsdale. 1986. An Anchoring and Adjustment Model of Spousal Predictions. *Journal of Consumer Research* 13, pp. 25–37.

Davis, Robert. 2006. "Wrong Site" Surgeries on the Rise. *USA Today*, April 18, p. D5.

Dawes, Robyn, David Faust, and Paul E. Meehl. 1989. Clinical Versus Actuarial Judgment. *Science* 243, pp. 1668–74.

Dean, Cornelia. 2005. Scientific Savvy? In U.S., Not Much. *New York Times*, Aug. 30.

Dean, John. 2007. Author correspondence.

Decina, Larry. 2007. Author interview.

DellaVigna, Stefano. 2007. Psychology and Economics: Evidence from the Field. Working paper.

———. 2008. Author interview.

DellaVigna, Stefano, and Ulrike Malmendier. 2004. Contract Design and Self-Control: Theory and Evidence. *Quarterly Journal of Economics* 119 (2), pp. 353–402.

———. 2005. Paying Not to Go to the Gym. *American Economic Review* 96 (3), pp. 694–719.

DellaVigna, Stefano, and Joshua M. Pollet. 2008. Investor Inattention and Friday Earnings Announcements. *Journal of Finance*, forthcoming.

Deloitte. 2007. Yes, Virginia, There Is a Santa Claus. Press release, Nov. 1.

DelVecchio, Devon, H. Shanker Krishnan, and Daniel C. Smith. 2007. Cents or Percent: The Effects of Promotion Framing on Price Expectations and Choice. *Journal of Marketing* 71 (3).

DePaulo, B. M., et al. 1996. Lying in Everyday Life. *Journal of Personality and Social Psychology* 70, pp. 979–95.

Diaz, Jaime. 1989. Perils of Putting. *Sports Illustrated,* April 3, pp. 76–79.

Diener, Ed, and Carol Diener. 1996. Most People Are Happy. *Psychological Science* 7 (3), pp. 181–85.

Dingus, T. A., et al. 2006. *The 100-Car Naturalistic Driving Study, Phase II—Results of the 100-Car Field Experiment.* National Highway Traffic Safety Administration, Report No. DOT HS 810 593, Washington, D.C.

DiVita, Joseph, et al. 2004. Verification of the Change Blindness Phenomenon While Managing Critical Events on a Combat Information Display. *Human Factors* 46 (20), pp. 205–18.

Doege, David. 2007. Man Mistakes Porn DVD as Woman's Cries for Help; He Faces Charges After Entering Apartment with Sword in Tow. *Milwaukee Journal Sentinel,* Feb. 20.

Doege, David, and Amy Rinard. 2007. Swordsman's Claim Disputed; Prosecutor Says Sounds on Neighbor's DVD Were Consensual Sex, Not Rape. *Milwaukee Journal Sentinel,* Feb. 21.

Donley, R., and M. Ashcraft. 1992. The Methodology of Testing Naïve Belief in the Physics Classroom. *Memory & Cognition* 20 (4), pp. 381–91.

Dooren, Jennifer Corbett. 2008. FDA Warns 2 Disodium Drugs Can Be Mistaken for Each Other. *Wall Street Journal,* Jan. 17, p. D4.

Douglas, Kate, et al. 2005. Attention Seeking. *New Scientist,* May 28, p. 38.

Dremen, David N., and Michael A. Berry. 1995. Analyst Forecasting Errors and Their Implications for Security Analysts. *Financial Analysts Journal,* May/June, pp. 30–41.

Dudokovich, Nicole, Elizabeth Marsh, and Barbara Tversky. 2004. Telling a Story or Telling It Straight: The Effects of Entertaining Versus Accurate Retellings on Memory. *Applied Cognitive Psychology* 18, pp. 125–43.

Dugas, Christine. 2007. Too Many 401(k)s Still Have Too Much Company Stock. *USA Today,* Dec. 14, 2007.

Duncker, K. 1939. The Influence of Past Experience upon Perceptual Properties. *American Journal of Psychology* 52, pp. 255–65.

Dunleavey, M. P. 2007. Disclosing Bias Doesn't Cancel Its Effects. *New York Times*, July 28, p. B6.

Dux, Paul E., Jason Ivanoff, Christopher L. Asplund, and Rene Marois. 2006. Isolation of a Central Bottleneck of Information Processing with Time-Resolved fMRI. *Neuron* 52, pp. 1109–20.

Dweck, Carol S. 1999. *Self-Theories: Their Role in Motivation, Personality, and Development.* Philadelphia: Psychology Press.

Ebbinghaus, Herman. 1964. *Memory: A Contribution to Experimental Psychology.* New York: Dover. (Originally published 1885; translated 1913.)

Einstein, Gilles, et al. 2003. Forgetting of Intentions in Demanding Situations Is Rapid. *Journal of Experimental Psychology: Applied* 9, pp. 147–62.

Einstein, Norman. 2007. Author interview.

Epley, Nicholas, and Erin Whitchurch. 2008. Mirror, Mirror on the Wall: Enhancement in Self-Recognition. *Personality and Social Psychology Bulletin* 34 (9), pp. 1159–70.

Ericsson, K. Anders. 1996. *The Road to Excellence: The Acquisition of Expert Performance in the Arts and Sciences, Sports and Games.* Mahwah, N.J.: Lawrence Erlbaum Associates.

———. 2007. Author interview.

Ericsson, K. Anders, Roy W. Roring, and Nandagopal Kiruthiga. 2007. Giftedness and Evidence for Reproducibly Superior Performance: An Account Based on the Expert Performance Framework. *High Ability Studies* 18 (1), pp. 3–56.

Ericsson, K. Anders, and Jacqui Smith, eds. 1991. *Toward a General Theory of Expertise: Prospects and Limits.* Cambridge, U.K.: Cambridge University Press.

Esch, H. E., and J. E. Burns. 1996. Distance Estimation by Foraging Honeybees. *Journal of Experimental Biology* 199 (1), pp. 155–62.

Etienne, Ariane S., et al. 1999. Dead Reckoning (Path Integration), Landmarks, and Representation of Space in a Comparative Perspective. In *Wayfinding Behavior: Cognitive Mapping and Other Spatial Processes*, edited by Reginald G. Golledge. Baltimore: Johns Hopkins University Press.

Fackelmann, Kathleen. 2006. Study: Long Hospital Shifts, Sleep Deprivation Can Kill. *USA Today*, Dec. 12, p. 1.

Feldman, R. S., J. A. Forrest, and B. R. Happ. 2002. Self-Presentation and Verbal Deception: Do Self-Presenters Lie More? *Basic and Applied Social Psychology* 24, pp. 163–70.

Ferraro, R., et al. 2005. Let Us Eat and Drink, for Tomorrow We Shall Die: Effects of Mortality Salience and Self-Esteem on Self-Regulation in Consumer Choice. *Journal of Consumer Research* 32, pp. 65–75.

Fielding, Rachel. 2003. Password Problems Swamp Help Desks. Vnunet.com, Jan. 15.

Fischhoff, Baruch. 1975. Hindsight ≠ Foresight: The Effect of Outcome Knowledge on Judgment Under Uncertainty. *Journal of Experimental Psychology: Human Perception and Performance* 1 (3), pp. 288–99.

———. 2007. An Early History of Hindsight Research. *Social Cognition* 25 (1), pp. 10–13.

Fischhoff, Baruch, and Ruth Beyth. 1975. "I Knew It Would Happen"— Remembered Probabilities of Once-Future Things. *Organizational Behavior and Human Performance* 13, pp. 1–16.

Fisher, Ian. 1998. In Excitement over Cancer Drugs, a Caution over Premature Hopes. *New York Times*, May 5.

Fontes, Miguel, and Peter Roach. 2007. Condom Nations. *Foreign Policy*, Sept./Oct.

Frank, Mark G., and Thomas Gilovich. 1988. The Dark Side of Self- and Social Perception: Black Uniforms and Aggression in Professional Sports. *Journal of Personality and Social Psychology* 54 (1), pp. 74–85.

Frank, Thomas. 2007. Most Fake Bombs Missed by Screeners. *USA Today*, Oct. 26.

Fryar, Cheryl D., et al. 2007. *Drug Use and Sexual Behaviors Reported by Adults: United States, 1999–2002*. Centers for Disease Control and Prevention, Advance Data No. 384, June 27.

Gaba, David M., et al. 2003. Differences in Safety Climate Between Hospital Personnel and Naval Aviators. *Human Factors* 45 (20), pp. 173–85.

Gabriel, Marsha T., Joseph W. Critelli, and Juliana S. Ee. 1994. Narcissistic Illusions in Self-Evaluations of Intelligence and Attractiveness. *Journal of Personality* 62 (1), pp. 143–55.

Galinsky, Adam D., and Thomas Mussweiler. 2001. First Offers as Anchors: The Role of Perspective-Taking and Negotiator Focus. *Journal of Personality and Social Psychology* 81 (4), pp. 657–69.

Gawande, Atul. 2007. The Power of Negative Thinking. *New York Times,* May 1.

Gervais, Simon, and Terrance Odean. 2001. Learning to Be Overconfident. *Review of Financial Studies* 14 (1), pp. 1–27.

Gibbons, Robert V., et al. 1998. A Comparison of Physicians' and Patients' Attitudes Toward Pharmaceutical Industry Gifts. *Journal of General Internal Medicine* 13, pp. 151–54.

Gibbs, W. Wayt. 2005. Considerate Computing. *Scientific American,* May 28, pp. 54–61.

Gibson, James J. 1950. *The Perception of the Visual World.* Boston: Houghton Mifflin.

———. 1979. *The Ecological Approach to Visual Perception.* Boston: Houghton Mifflin.

Gilbert, Daniel. 2005. *Stumbling on Happiness.* New York: Knopf.

———. 2006. I'm O.K., You're Biased. *New York Times,* April 16, p. WK12.

Gilbert, Daniel, and Jane E. Ebert. 2002. Decisions and Revisions: The Affective Forecasting of Changeable Outcomes. *Journal of Personality and Social Psychology* 82 (4), pp. 503–14.

Gilbert, Daniel, et al. 1998. Immune Neglect: A Source of Durability Bias in Affective Forecasting. *Journal of Personality and Social Psychology* 75 (3), pp. 617–38.

Gilovich, Thomas. 1981. Seeing the Past in the Present: The Effect of Associations to Familiar Events on Judgments and Decisions. *Journal of Personality and Social Psychology* 40 (5), pp. 797–808.

———. 1983. Biased Evaluation and Persistence in Gambling. *Journal of Personality and Social Psychology* 44 (6), pp. 1110–26.

———. 2007. Author interview.

Gilovich, Thomas, Victoria Husted Medvec, and Serena Chen. 1995. Commission, Omission, and Dissonance Reduction: Coping with Regret in the "Monty Hall" Problem. *Personality and Social Psychology Bulletin* 21 (2), pp. 182–90.

Glenberg, Arthur M., and William Epstein. 1985. Calibration of Comprehension. *Journal of Experimental Psychology: Learning, Memory, and Cognition* 11 (4), pp. 702–18.

Godden, D. R., and A. D. Baddeley. 1975. Context-Dependent Memory in Two Natural Environments: On Land and Underwater. *British Journal of Psychology* 66 (3), pp. 325–31.

Goldberg, Carey. 2005. If You Don't Find It Often, You Don't Often Find It. *Boston Globe*, May 31, p. C1.

Goldberg, Lewis R. 1968. Simple Models or Simple Processes? Some Research on Clinical Judgments. *American Psychologist* 23, pp. 483–96.

Golledge, Reginald G., ed. 1999. *Wayfinding Behavior: Cognitive Mapping and Other Spatial Processes*. Baltimore: Johns Hopkins University Press.

Gonzalez, Victor, and Gloria Mark. 2004. "Constant, Constant, Multi-tasking Craziness": Managing Multiple Working Spheres. *2004 Proceedings of Human Factors in Computer Systems*, Vienna, 6 (1).

Gowen, Annie, and Lila Arzua. 2004. 10 Teens Hurt as Tour Bus Slams into Overpass. *Washington Post*, Nov. 15, p. B1.

Gramzow, Richard H., Greg Willard, and Wendy Berry Mendes. 2008. Big Tales and Cool Heads: Academic Exaggeration Is Related to Cardiac Vagal Reactivity. *Emotion* 8 (1), pp. 138–44.

Green, Paul. 1999. Visual and Task Demands of Driver Information Systems. Technical report UMTRI-98-16. Ann Arbor: University of Michigan Transportation Research Institute.

————. 2004. Driver Distraction, Telematics Design, and Workload Managers: Safety Issues and Solutions. Convergence Transportation Electronics Association.

Greenberg, Richard N. 1984. Overview of Patient Compliance with Medication Dosing: A Literature Review. *Clinical Therapeutics* 6 (5), pp. 592–99.

Greene, Terry, and Helga Noice. 1988. Influence of Positive Affect upon Creative Thinking and Problem Solving in Children. *Psychological Reports* 63, pp. 895–98.

Grinblatt, Mark, and Matti Keloharju. 2008. Sensation Seeking, Overconfidence, and Trading Activity. Working paper available at www.anderson.ucla.edu/documents/areas/fac/finance/06-06.pdf.

Grow, Brian. 2005. The Great Rebate Runaround. *BusinessWeek*, Nov. 23.

Gruneberg, M. M., P. E. Morris, and R. N. Sykes, eds. 1978. *Practical Aspects of Memory*. London: Academic Press.

Hagerty, James R., and Kemba J. Dunham. 2005. How Big U.S. Home Builders Plan to Ride out a Downturn. *Wall Street Journal Online*, Dec. 1.

Hallenbeck, Cleve. 1920. Forecasting Precipitation in Percentages of Probability. *Monthly Weather Review* 48 (11), pp. 645–47.

Hallinan, Joseph T. 2005. Heal Thyself: Once Seen as Risky, One Group of Doctors Changes Its Ways. *Wall Street Journal*, June 21, p. A1.

———. 2007. Investor Appetite Is What Diet Firm May Really Lose. *Wall Street Journal*, Feb. 1, p. C1.

Hancock, H., A. Fisk, and W. Rogers. 2005. Comprehending Product Warning Information: Age-Related Effects and the Roles of Memory, Inferencing, and Knowledge. *Human Factors* 47 (2), pp. 219–34.

Harbluk, Joanne L., Y. Ian Noy, and Moshe Eizenman. 2002. The Impact of Cognitive Distraction on Driver Visual Behaviour and Vehicle Control. *Transport Canada*, TP No. 13889 E, Feb.

*Harper's*. 2007. Harper's Index. July, p. 15.

Harris, Christine R., Michael Jenkins, and Dale Glaser. 2006. Gender Differences in Risk Assessment: Why Do Women Take Fewer Risks Than Men? *Judgment and Decision Making* 1 (1), pp. 48–63.

Harris, Gardiner. 2008. Drug Industry to Announce Revised Code on Marketing. *New York Times,* July 10.

Harrison, Yvonne, and James A. Horne. 2000. The Impact of Sleep Deprivation on Decision Making: A Review. *Journal of Experimental Psychology: Applied* 6 (3), pp. 236–49.

Haslam, Diana R., and Peter Abraham. 1987. Sleep Loss and Military Performance. In *Contemporary Studies in Combat Psychiatry,* edited by G. Belenky. Westport, Conn.: Greenwood Press.

Hastie, Reid, David A. Schkade, and John W. Payne. 1999. Juror Judgments in Civil Cases: Effects of Plaintiff's Requests and Plaintiff's Identity on Punitive Damage Awards. *Law and Human Behavior* 23 (4), pp. 445–70.

Held, Richard, et al. 2008. Revisiting the Molyneux Question. *Journal of Vision* 8 (6), pp. 523, 523a.

Herman, C. Peter, and Janet Polivy. 2003. Dieting as an Exercise in

Behavioral Economics. In *Time and Decision*, edited by Daniel Read. New York: Russell Sage Foundation.

Hertwig, Ralph, Thorsten Pachur, and Stephanie Kurzenhauser. 2005. Judgments of Risk Frequencies: Tests of Possible Cognitive Mechanisms. *Journal of Experimental Psychology: Learning, Memory, and Cognition* 31 (4), pp. 621–42.

Heuer, R. 1999. *Psychology of Intelligence Analysis.* Center for the Study of Intelligence, Central Intelligence Agency, Washington D.C. Available at www.cia.gov/library/.

Hewitt, Claire. 2007. Author interview.

Heywood, Simon. 1972. The Popular Number Seven or Number Preference. *Perceptual and Motor Skills* 34, pp. 357–58.

Hill, Michael. 2006. Professor Sniffs Out Folks' Eating Habits. Associated Press, Nov. 6.

Ho, Daniel E., and Kosuke Imai. 2006. Estimating Causal Effects of Ballot Order from a Randomized Natural Experiment: California Alphabet Lottery, 1978–2002. Working paper, draft as of Nov. 18.

Holland, C., and P. Rabbitt. 1992. People's Awareness of Their Age-Related Sensory and Cognitive Deficits and the Implications for Road Safety. *Applied Cognitive Psychology* 6, pp. 217–31.

Hollnagel, Erik. 1993. *Human Reliability Analysis Context and Control.* London: Academic Press.

Huberman, Gur, and Tomer Regev. 2001. Contagious Speculation and a Cure for Cancer: A Nonevent That Made Stock Prices Soar. *Journal of Finance* 56 (1), pp. 387–96.

Hull, A., A. Wilkins, and A. Baddeley. 1988. Cognitive Psychology and the Wiring of Plugs. In *Practical Aspects of Memory: Current Research and Issues.* Vol. 1: *Memory in Everyday Life*, edited by M. M. Gruneberg, P. E. Morris, and R. N. Sykes. Chichester, U.K.: Wiley.

Inhoff, Albrecht Werner, Robin Morris, and John Calabrese. 1986. Eye Movements in Skilled Transcription Typing. *Bulletin of the Psychonomic Society* 24 (2), pp. 113–14.

Isen, Alice M. 2001. An Influence of Positive Affect on Decision Making in Complex Situations: Theoretical Issues with Practical Implications. *Journal of Consumer Psychology* 11 (2), pp. 75–85.

Isen, Alice M., K. Daubman, and G. Nowicki. 1987. Positive Affect

Facilitates Creative Problem Solving. *Journal of Personality and Social Psychology* 62 (6), pp. 1122–31.

Isen, Alice M., Thomas E. Nygren, and F. Gregory Ashby. 1988. Influence of Positive Affect on Subjective Utility of Gains and Losses: It Is Just Not Worth the Risk. *Journal of Personality and Social Psychology* 55 (5), pp. 710–17.

Isen, Alice M., and Robert Patrick. 1983. The Effect of Positive Feeling on Risk Taking: When the Chips Are Down. *Organizational Behavior and Human Performance* 31, pp. 194–202.

Janelle, Christopher M., and Charles H. Hillman. 2003. Expert Performance in Sport: Current Perspectives and Critical Issues. In *Expert Performance in Sports: Advances in Research on Sport Expertise*, edited by Janet L. Starkes and K. Anders Ericsson. Champaign, Ill.: Human Kinetics.

Jargon, Julie, Ann Zimmerman, and David Kesmodel. 2008. Grocers Tout "Sales" Even as Prices Climb. *Wall Street Journal*, April 1, p. B1.

Jiang, Yuhong. 2004. Resolving Dual-Task Interference: An fMRI Study. *NeuroImage* 22, pp. 748–54.

Jiang, Yuhong, Rebecca Saxe, and Nancy Kanwisher. 2004. Functional Magnetic Resonance Imaging Provides New Constraints on Theories of the Psychological Refractory Period. *Psychological Science* 15 (6), pp. 390–96.

Johnson, D. P., et al. 2006. Overconfidence in Wargames: Experimental Evidence on Expectations, Aggression, Gender, and Testosterone. *Proceedings of the Royal Society B* 273, pp. 2513–20.

Johnson, M. H., et al. 1991. Newborns' Preferential Tracking of Face-Like Stimuli and Its Subsequent Decline. *Cognition* 40, pp. 1–19.

Johnson, Richard F., and Donna J. Merullo. 1999. Friend-Foe Discrimination, Caffeine, and Sentry Duty. *Proceedings of the Human Factors and Ergonomics Society 43rd Annual Meeting*, pp. 1348–52.

Johnson-Laird, Philip N. 2006. *How We Reason*. New York: Oxford University Press.

Johnson-Laird, Philip N., and P. C. Wason, eds. 1977. *Thinking: Readings in Cognitive Science*. Cambridge, U.K.: Cambridge University Press.

Jones, Gregg, Holly Becka, Jennifer LaFleur, and Steve McGonigle. 2006. After Lives Are Lost, Records Often Go Missing, *Dallas Morning News*, Sept. 17.

Jones, Gregory V. 1990. Misremembering a Common Object: When Left Is Not Right. *Memory & Cognition* 18 (2), pp. 174–82.

Kahneman, Daniel, et al. 2004. A Survey Method for Characterizing Daily Life Experience: The Day Reconstruction Method. *Science* 306, pp. 1776–80.

———. 2006. Would You Be Happier If You Were Richer? A Focusing Illusion. *Science* 312, pp. 1908–10.

Kahneman, Daniel, and W. Scott Peavler. 1969. Incentive Effects and Pupillary Changes in Association Learning. *Journal of Experimental Psychology* 79, pp. 312–18.

Kahneman, Daniel, and Amos Tversky. 1973. On the Psychology of Prediction. *Psychological Review* 80 (4), pp. 237–51.

———. 1979. Prospect Theory: An Analysis of Decision Under Risk. *Econometrica* 47 (2), pp. 363–91.

Kendall, Peter. 1994. PINS Pile Up Until They Become Personal Pain. *Chicago Tribune*, Oct. 14, p. C1.

Keren, Gideon. 1987. Facing Uncertainty in the Game of Bridge: A Calibration Study. *Organizational Behavior and Human Decision Making* 39, pp. 98–114.

Keysar, Boaz, and Anne S. Henley. 2002. Speakers' Overestimation of Their Effectiveness. *Psychological Science* 13 (3), pp. 207–12.

Kiley, David, and Earle Eldridge. 2002. Are Those Gizmos on the Dashboard Too Distracting? *USA Today*, May 29.

Killion, Ann. 2007. Author interview.

King, Gerhart F. 1986. Driver Performance in Highway Navigation Tasks. *Transportation Research Record* 1093, pp. 1–11.

Klauer, S. G. 2007. Author interview.

Klauer, S. G., et al. 2006. *The Impact of Driver Inattention on Near-Crash/Crash Risk: An Analysis Using the 100-Car Naturalistic Driving Study*. National Highway Traffic Safety Administration, Report No. DOT HS 810 594, Washington, D.C.

Klein, Stefan. 2007. *The Secret Pulse of Time*. New York: Marlowe.

Klein, William M., and Ziva Kunda. 1993. Maintaining Self-Serving Social Comparisons: Biased Reconstruction of One's Past Behaviors. *Personality and Social Psychology Bulletin* 19, pp. 732–39.

Kluger, Jeffrey. 2006. Why We Worry About the Things We Shouldn't ... and Ignore the Things We Should. *Time*, Dec. 4, pp. 64–71.

Konigsberg, Eric. 2008. City Council Bill Would Ban Text Messaging While Driving. *New York Times*, Aug. 15, p. A16.

Kruger, Justin. 2007. Author interview.

Kruger, Justin, Derrick Wirtz, and Dale T. Miller. 2005. Counterfactual Thinking and the First Instinct Fallacy. *Journal of Personality and Social Psychology* 88, pp. 725–35.

Kubovy, Michael. 1977. Response Availability and the Apparent Spontancity of Numerical Choices. *Journal of Experimental Psychology: Human Perception and Performance* 3 (2), pp. 359–64.

Kubovy, Michael, and Joseph Psotka. 1976. The Predominance of Seven and the Apparent Spontaneity of Numerical Choices. *Journal of Experimental Psychology: Human Perception and Performance* 2 (2), pp. 291–94.

Kuebli, Janet, and Robyn Fivush. 1992. Gender Differences in Parent-Child Conversations About Past Emotions. *Sex Roles* 27 (11/12), pp. 683–98.

Lambert, Craig. 2006. The Marketplace of Perceptions: Behavioral Economics Explains Why We Procrastinate, Buy, Borrow, and Grab Chocolate on the Spur of the Moment. *Harvard Magazine*, March–April, p. 50.

Landro, Laura. 2008. Hospitals Tackle High-Risk Drugs to Reduce Errors. *Wall Street Journal*, March 5, p. D1.

Landy, David, and Harold Sigall. 1974. Beauty Is Talent: Task Evaluation as a Function of the Performer's Physical Attractiveness. *Journal of Personality and Social Psychology* 29 (3), pp. 299–304.

Langer, Ellen. 1975. The Illusion of Control. *Journal of Personality and Social Psychology* 32 (2), pp. 311–28.

Langer, Ellen, and Jane Roth. 1975. Heads I Win, Tails It's Chance: The Illusion of Control as a Function of the Sequence of Outcomes in a Purely Chance Task. *Journal of Personality and Social Psychology* 32 (6), pp. 951–55.

Lawton, Carol A. 1994. Gender Differences in Way-Finding Strategies: Relationship to Spatial Ability and Spatial Anxiety. *Sex Roles* 30 (11/12), pp. 765–79.

———. 2001. Gender and Regional Differences in Spatial Referents Used in Direction Giving. *Sex Roles* 44 (5/6), pp. 321–37.

Lawton, Carol A., and Janos Kallai. 2002. Gender Differences in Wayfinding Strategies and Anxieties About Wayfinding: A Cross-Cultural Comparison. *Sex Roles* 47 (9/10), pp. 389–401.

Leibovich, Mark. 2004. George Tenet's "Slam-Dunk" into the History Books. *Washington Post*, June 4, p. C1.

Leonhardt, David. 2006. Why Doctors So Often Get It Wrong. *New York Times*, Feb. 22, p. C1.

Levin, Alan, and Brad Heath. 2007. Fatigue Key to Air Crew Errors. *USA Today*, Nov. 8, p. 1.

Levin, Daniel T., and Daniel J. Simons. 1997. Failure to Detect Changes to Attended Objects in Motion Pictures. *Psychonomic Bulletin & Review* 4 (4), pp. 501–6.

Lichtenstein, Sarah, and Baruch Fischhoff. 1977. Do Those Who Know More Also Know More About How Much They Know? *Organizational Behavior and Human Performance* 20, pp. 159–83.

Lichtenstein, Sarah, Baruch Fischhoff, and Lawrence D. Phillips. 1982. Calibration of Probabilities: The State of the Art to 1980. In *Judgment Under Uncertainty*, edited by Daniel Kahneman et al. Cambridge, U.K.: Cambridge University Press.

Linderman, Lawrence. 1979. Playboy Interview: Burt Reynolds. *Playboy* 26 (10), pp. 67–94.

Lobben, Amy K. 2004. Tasks, Strategies, and Cognitive Processes Associated with Navigational Map Reading: A Review Perspective. *Professional Geographer* 56 (2), pp. 270–81.

Loewenstein, George. 2007. Author interview.

Loewenstein, George, and David Schkade. 1999. Wouldn't It Be Nice? Predicting Future Feelings. In *Well-Being: The Foundations of Hedonic Psychology*, edited by Daniel Kahneman, Ed Diener, and Norbert Schwarz. New York: Russell Sage Foundation.

Loftus, Elizabeth, et al. 1987. Time Went By So Slowly: Overestimation of Event Duration by Males and Females. *Applied Cognitive Psychology* 1, pp. 3–13.

Lohr, Steve. 2007. Slow Down, Brave Multitasker, and Don't Read This in Traffic. *New York Times*, March 25.

Loomis, Jack M., Roberta L. Klatzky, Reginald G. Golledge, and John W.

Philbeck. 1999. Human Navigation by Path Integration. In *Wayfinding Behavior: Cognitive Mapping and Other Spatial Processes,* edited by Reginald G. Golledge. Baltimore: Johns Hopkins University Press.

Lord, Albert B. 1960. *The Singer of Tales.* Cambridge, Mass.: Harvard University Press.

Lorentz, G. B. A., et al. 1999. Miss Rate of Lung Cancer on the Chest Radiograph in Clinical Practice. *Chest* 115, pp. 720–24.

Luchins, A. S. 1942. Mechanization in Problem Solving. *Psychological Monograms* 54 (248), pp. 1–95.

Luchins, A. S., and E. H. Luchins. 1950. New Experimental Attempts at Preventing Mechanization in Problem Solving. *Journal of General Psychology* 42, pp. 279–97.

Lundberg, George D. 1998. Low-Tech Autopsies in the Era of High-Tech Medicine. *JAMA* 280 (14), pp. 1273–74.

Lykken, David, and Auke Tellegen. 1996. Happiness Is a Stochastic Phenomenon. *Psychological Science* 7 (3), pp. 186–89.

Maccoby, Eleanor Emmons, and Carol Nagy Jacklin. 1974. *The Psychology of Sex Differences.* Stanford, Calif.: Stanford University Press.

Mack, Arien, and Irvin Rock. 1998. Inattentional Blindness. Cambridge, Mass.: MIT Press.

Macur, Juliet. 2007. Rule Jostles Runners Who Race to Their Own Tune. *New York Times,* Nov. 1, p. 1.

Malmendier, Ulrike, and Geoffrey Tate. 2005. CEO Overconfidence and Corporate Investment. *Journal of Finance* 60 (6), pp. 2661–700.

Manhart, Klaus. 2004. The Limits of Multitasking. *Scientific American Mind,* Dec.

Marcus, Erin N. 2006. When Young Doctors Strut Too Much of Their Stuff. *New York Times,* Nov. 21.

Martin, M., and Jones, G. 1999. Hale-Bopp and Handedness: Individual Differences in Memory for Orientation. *Psychological Science* 10 (3), pp. 267–69.

Martinez-Conde, Susana, and Macknik, Stephen L. 2007. Windows on the Mind. *Scientific American,* Aug., pp. 56–63.

Mateja, Jim. 2007. S80 Makes Big Impression in a Flash. *Chicago Tribune,* March 25.

Mathews, C. O. 1929. Erroneous First Impressions on Objective Tests. *Journal of Educational Psychology* 20, pp. 280–86.

Mathur, P., and J. Kruger. 2007. The First Instinct Fallacy Among Investors. Unpublished data.

Matthews, M. H. 1987. Gender, Home Range, and Environmental Cognition. *Transactions of the Institute of British Geographers,* n.s., 12 (1), pp. 43–56.

————. 1992. *Making Sense of Place: Children's Understanding of Large-Scale Environments.* Hertfordshire, U.K.: Harvester Wheatsheaf.

Matthews, S. 1997. Proposals for Improving Aviation Safety and Changing the System. Remarks to the White House Commission on Aviation Safety and Security. International Conference on Aviation Safety and Security in the Twenty-first Century, Washington, D.C., Jan. 13.

May, Andrew J., and Tracy Ross. 2006. Presence and Quality of Navigational Landmarks: Effect on Driver Performance and Implications for Design. *Human Factors* 48 (2), pp. 346–61.

Mayer, Caroline E. 2002. Why Won't We Read the Manual? *Washington Post,* May 26, p. H01.

Maynard, Micheline. 2007. At Chrysler, Home Depot Still Lingers. *New York Times,* Oct. 30.

McCartney, Scott. 2005. Addressing Small Errors in the Cockpit. *Wall Street Journal,* Sept. 13, p. D1.

————. 2006. When Pilots Pass the BRBON, They Must Be in Kentucky. *Wall Street Journal,* March 21, p. A1.

McCloskey, Michael. 1983. Intuitive Physics. *Scientific American* 248 (4), pp. 122–30.

McCloskey, Michael, Allyson Washburn, and Linda Felch. 1983. Intuitive Physics: The Straight-Down Belief and Its Origin. *Journal of Experimental Psychology: Learning, Memory, and Cognition* 9 (4), pp. 636–49.

McConnell, Steve. 1996. *Rapid Development.* Redmond, Wash.: Microsoft Press.

McCracken, Vicki A. 2007. Author interview.

McCracken, Vicki A., Robert D. Boynton, and Brian F. Blake. 1982. The Impact of Comparative Food Price Information on Consumers and Grocery Retailers: Some Preliminary Findings of a Field Experiment. *Journal of Consumer Affairs* 16 (2), pp. 224–40.

McGlothlin, William H. 1956. Stability of Choices Among Uncertain Alternatives. *American Journal of Psychology* 69, pp. 604–15.

Meiners, Jens. 2004. BMW Sticks to iDrive; System Will Be Simplified but Is Destined to Stay. *Automotive News,* Oct.

Menand, Louis. 2005. Everybody's an Expert. *New Yorker,* Dec. 5, pp. 98–101.

Merrick, Amy. 2008. Limited Brands Expects Troubles to Continue. *Wall Street Journal,* Feb. 28, p. C6.

Merriweather, James. 2005. Woman Found Hanging on Tree. *Wilmington* (Del.) *News Journal,* Oct. 27, p. B4B.

Middlebrooks, S. E., B. G. Knapp, and B. W. Tillman. 1999. *An Evaluation of Skills and Abilities Required in the Simultaneous Performance of Using a Mobile Telephone and Driving an Automobile.* U.S. Army Research Laboratory, Aberdeen Proving Ground, Md.

Milgram, Stanley. 1974. *Obedience to Authority.* New York: Harper & Row.

Milgram, Stanley, and Denise Jodelet. 1976. Psychological Maps of Paris. In *Environmental Psychology: People and Their Physical Settings.* 2nd ed., edited by Harold M. Proshansky, William H. Ittelson, and Leanne G. Rivlin. New York: Holt, Rinehart, and Winston.

Miller, Geoffrey, Joshua M. Tybur, and Brent D. Jordan. 2007. Ovulatory Cycle Effects on Tip Earnings by Lap Dancers: Economic Evidence for Human Estrus? *Evolution and Human Behavior* 28, pp. 375–81.

Miller, Stephen. 2006. Remembrances: Father of Auto Rebate Changed Car Buying in U.S. *Wall Street Journal,* Nov. 18, p. A6.

Mindlin, Alex. 2006. For a Memorable Price, Trim the Syllables. *New York Times,* Aug. 14, p. C3.

———. 2007. It's How Much You Think You Have. *New York Times,* Nov. 26, p. C3.

Mocan, Naci, and Erdal Tekin. 2006. Ugly Criminals. Working paper 12019, National Bureau of Economic Research.

Monin, Benoit, and Dale T. Miller. 2001. Moral Credentials and the Expression of Prejudice. *Journal of Personality and Social Psychology* 81 (1), pp. 33–43.

Monk, Christopher A., Deborah A. Boehm-Davis, and J. Gregory Trafton. 2004. Recovering from Interruptions: Implications for Driver Distraction Research. *Human Factors* 46 (4), pp. 650–63.

Montello, Daniel R. 2007. Author interview.

Montello, Daniel R., Kristin L. Lovelace, Reginald G. Golledge, and Carole M. Self. 1999. Sex-Related Differences and Similarities in Geographic and Environmental Spatial Abilities. *Annals of the Association of American Geographers* 89 (3), pp. 515–34.

Montepare, J. M., and L. A. Zebrowitz. 1998. Person Perception Comes of Age: The Salience of Age in Social Judgments. *Advances in Experimental Social Psychology* 30, p. 93.

Mook, Douglas. 2004. *Classic Experiments in Psychology.* Westport, Conn.: Greenwood Press.

*Morbidity and Mortality Weekly Report.* 2007. Nail-Gun Injuries Treated in Emergency Departments—United States, 2001–2005. 56 (14), pp. 329–32.

Morewedge, Carey K., Leif Holtzman, and Nicholas Epley. 2007. Unfixed Resources: Perceived Costs, Consumption, and the Accessible Account Effect. *Journal of Consumer Research* 34, pp. 459–67.

Morin, Rich. 2006. The Ugly Face of Crime. *Washington Post,* Feb. 17, p. A2.

Moroze, Michael L., and Michael P. Snow. 1999. *Causes and Remedies of Controlled Flight into Terrain in Military and Civil Aviation.* Air Force Research Laboratory, Wright-Patterson Air Force Base, Dayton, Ohio.

Mueller, Ulrich, and Allan Mazur. 1996. Facial Dominance of West Point Cadets as a Predictor of Later Military Rank. *Social Forces* 74 (3), pp. 823–50.

Muhm, John R., et al. 1983. Lung Cancer Detected During a Screening Program Using Four-Month Chest Radiographs. *Radiology* 148, pp. 609–15.

Murphy, Allan H., and Robert L. Winkler. 1984. Probability Forecasting in Meteorology. *Journal of the American Statistical Association* 79, pp. 489–500.

Mussweiler, Thomas, Fritz Strack, and Tim Pfeiffer. 2000. Overcoming the Inevitable Anchoring Effect: Considering the Opposite Compensates for Selective Accessibility. *Personality and Social Psychology Bulletin* 26 (9), pp. 1142–50.

National Highway Traffic Safety Administration. 1986. *An Evaluation of Child Passenger Safety: The Effectiveness and Benefits of Safety Seats.* DOT HS 806 890, Feb.

————. 1994. *Examination of Lane Change Crashes and Potential IVHS Countermeasures.* DOT HS 808 071, March.

————. 2004. *Misuse of Child Restraints.* DOT HS 809 671, Jan.

————. 2006. *Child Restraint Use Survey.* DOT HS 810 679, Dec.

National Institute for Health Care Management. 2002. *Changing Patterns of Pharmaceutical Innovation.* Washington, D.C.

National Transportation Safety Board. 1973. *Aircraft Accident Report.* Report No. NTSB-AAR-73-14. Adopted June 14.

————. 1979. *Aircraft Accident Report.* Report No. NTSB-AAR-79-7. Adopted June 7.

————. 1992. *Aircraft Accident Report.* PB92-910402, Report No. NTSB/AAR 920-02. Adopted March 18.

————. 2006a. *Motorcoach Collision with the Alexandria Avenue Bridge Overpass, George Washington Memorial Parkway, Alexandria, Va., Nov. 14, 2004.* Highway Accident Report NTSB/HAR-06/04, Washington, D.C.

————. 2006b. *Air Accident Brief.* Accident No. IAD05FA023. Adopted Nov. 8.

Neikirk, William. 2007. States Told to Prep for Gray Driver Boom. *Chicago Tribune,* April 12, p. 3.

Neisser, Ulric. 1981. John Dean's Memory: A Case Study. *Cognition* 9, pp. 1–22.

————. 1988. Time Present and Past. In *Practical Aspects of Memory: Current Research and Issues.* Vol. 2: *Clinical and Educational Implications,* edited by M. M. Gruneberg, P. E. Morris, and R. N. Sykes. Chichester, U.K.: Wiley.

————. 2007. Author interview.

Neisser, Ulric, and Ira E. Hyman Jr. 2000. *Memory Observed: Remembering in Natural Contexts.* 2nd ed. New York: Worth.

Neville, Kelly J., et al. 1994. Subjective Fatigue of C-141 Aircrews During Operation Desert Storm. *Human Factors* 36 (2), pp. 339–49.

Newby-Clark, Ian R., et al. 2000. People Focus on Optimistic Scenarios and Disregard Pessimistic Scenarios While Predicting Task Completion Times. *Journal of Experimental Psychology: Applied* 6 (3), pp. 171–82.

Newport, John Paul. 2007. The Eyes Have It. *Wall Street Journal,* Oct. 27–28, p. W1.

Nickerson, Raymond S., and Marilyn Jager Adams. 1979. Long-Term Memory for a Common Object. *Cognitive Psychology* 11, pp. 287–307.

Norman, D. A. 1976. *Memory and Attention.* New York: Wiley.

———. 1988. *The Design of Everyday Things.* New York: Basic Books.

———. 1992. *Turn Signals Are the Facial Expressions of Automobiles.* Cambridge, Mass.: Perseus.

North, Adrian C., David J. Hargreaves, and Jennifer McKendrick. 1997. In-Store Music Affects Product Choice. *Nature* 390, p. 132.

Northcraft, Gregory B., and Margaret A. Neale. 1987. Experts, Amateurs, and Real Estate: An Anchoring-and-Adjustment Perspective on Property Pricing Decisions. *Organizational Behavior and Human Decision Processes* 39, pp. 84–97.

Odean, Terrance. 1999. Do Investors Trade Too Much? *American Economic Review* 89 (5), pp. 1279–98.

Ornstein, Charles. 2007. Dennis Quaid Files Suit over Drug Mishap. *Los Angeles Times,* Dec. 5.

———. 2008. Quaids Recall Twins' Drug Overdose. *Los Angeles Times,* Jan. 15.

Oskamp, Stuart. 1965. Overconfidence in Case-Study Judgments. *Journal of Consulting Psychology* 29 (3), pp. 261–65.

Paese, Paul W., and Janet A. Sniezek. 1991. Influences on the Appropriateness of Confidence in Judgment: Practice, Effort, Information, and Decision-Making. *Organizational Behavior and Human Decision Processes* 48, pp. 100–130.

Parfitt, Tom. 2007. Revealed: Why Those Russian Submarine Heroics Might Have Looked a Little Familiar. *Guardian,* Aug. 11.

Pashler, Harold. 1994. Dual-Task Interference in Simple Tasks: Data and Theory. *Psychological Bulletin* 116 (2), pp. 220–44.

Pasztor, Andy. 2007. Continental, American Balk at New Rest Rules for International-Flight Crews. *Wall Street Journal,* March 20, p. A8.

Pasztor, Andy, and Susan Carey. 2006. Pilot-Fatigue Test Lands JetBlue in Hot Water. *Wall Street Journal,* Oct. 21, p. A1.

Paumgarten, Nick. 2006. The $40 Million Elbow. *New Yorker,* Oct. 23.

Payne, J. W., J. R. Bettman, and E. J. Johnson. 1993. *The Adaptive Decision Maker.* Cambridge, U.K.: Cambridge University Press.

Pearce, Craig L. 2008. Follow the Leaders: You've Created a Team to Solve a Problem. Here's Some Advice: Don't Put One Person in Charge. *Wall Street Journal,* July 7, p. R8.

Peirce, Sean, and Jane Lappin. 2006. *Private Sector Deployment of Intelligent Transportation Systems: Current Status and Trends.* U.S. Department of Transportation, Research and Innovative Technology Administration, Cambridge, Mass.

Peterson, Carole, and Regina Rideout. 1998. Memory for Medical Emergencies Experienced by 1- and 2-Year-Olds. *Developmental Psychology* 34 (5), pp. 1059–72.

Pilcher, June J., and Allen I. Huffcutt. 1996. Effects of Sleep Deprivation on Performance: A Meta-analysis. *Sleep* 19 (4), pp. 318–26.

Pipitone, R., and G. Gallup. 2008. Women's Voice Attractiveness Varies Across the Menstrual Cycle. *Evolution and Human Behavior,* online, April.

Plassman, Hilke, John O'Doherty, Baba Shiv, and Antonio Rangel. 2008. Marketing Actions Can Modulate Neural Representations of Experienced Pleasantness. *Proceedings of the National Academy of Sciences* 105 (3), pp. 1050–54.

Plous, S. 1989. Thinking the Unthinkable: The Effects of Anchoring on Likelihood Estimates of Nuclear War. *Journal of Applied Social Psychology* 19, pp. 67–91.

Pollock, Andrew. 2004. With New Sleeping Pill, New Acceptability? *New York Times,* Dec. 17.

Possley, Maurice. 2006. "I Have to Make This Right." *Chicago Tribune,* Nov. 15, p. 1.

Powers, P. A., J. L. Andriks, and E. F. Loftus. 1979. Eyewitness Accounts of Females and Males. *Journal of Applied Psychology* 64, pp. 339–47.

Pradhan, Anuj, et al. 2005. Using Eye Movements to Evaluate Effects of Driver Age on Risk Perception in a Driving Simulator. *Human Factors* 47 (4), pp. 840–52.

Prinsell, C. P., P. H. Ramsey, and P. P. Ramsey. 1994. Score Gains, Attitudes, and Behavior Changes due to Answer-Changing Instruction. *Journal of Educational Measurement* 31 (4), pp. 327–37.

Raghavan, Anita. 2004. Surely They Jest. *Wall Street Journal,* Aug. 9, p. A1.

Read, Daniel. 2005. Monetary Incentives, What Are They Good For? *Journal of Economic Methodology* 12 (2), pp. 265–76.

Read, Daniel, George Loewenstein, and Shobana Kalyanaraman. 1999. Mixing Virtue and Vice: The Combined Effects of Hyperbolic Discounting and Diversification. *Journal of Behavioral Decision Making* 12, pp. 257–73.

Read, Daniel, and Barbara van Leeuwen. 1998. Predicting Hunger: The Effects of Appetite and Delay on Choice. *Organizational Behavior and Human Decision Processes* 76 (2), pp. 189–205.

Reason, James. 1990. *Human Error.* Cambridge, U.K.: Cambridge University Press.

Recarte, Miguel A., and Luis M. Nunes. 2000. Effects of Verbal and Spatial-Imagery Tasks on Eye Fixations While Driving. *Journal of Experimental Psychology: Applied* 6 (1), pp. 31–43.

———. 2003. Mental Workload While Driving: Effects on Visual Search, Discrimination, and Decision Making. *Journal of Experimental Psychology: Applied* 9 (2), pp. 119–37.

Redelmeier, Donald A., and Robert J. Tibshirani. 1999. Why Cars in the Next Lane Seem to Go Faster. *Nature* 401, p. 35.

———. 2001. Car Phones and Car Crashes: Some Popular Misconceptions. *Canadian Medical Association Journal* 164 (11).

Reder, Lynne M., and John R. Anderson. 1980. A Comparison of Texts and Their Summaries: Memorial Consequences. *Journal of Verbal Learning and Verbal Behavior* 19, pp. 121–34.

Reilly, Jacqueline, and Gerry Mulhern. 1995. Gender Differences in Self-Estimated IQ: The Need for Care in Interpreting Group Data. *Personality and Individual Differences* 18 (2), pp. 189–92.

Rensink, Ronald A., Kevin O'Regan, and James J. Clark. 1997. To See or Not to See: The Need for Attention to Perceive Changes in Scenes. *Psychological Science* 8 (5), pp. 368–73.

Reuters. 2007a. Hey, Big Boy! Any Interest? Sept. 13.

———. 2007b. Remember Your Home Phone Number? Forget It! July 7.

Richard, Christian M., et al. 2002. Effect of a Concurrent Auditory Task on Visual Search Performance in a Driving-Related Image-Flicker Task. *Human Factors* 44 (1), pp. 108–19.

Robbins, Lillian C. 1963. The Accuracy of Parental Recall of Aspects of Child Development and Child Rearing Practices. *Journal of Abnormal and Social Psychology* 66, pp. 261–70.

Roberts, S. Craig, et al. 2004. Female Facial Attractiveness Increases During the Fertile Phase of the Menstrual Cycle. *Proceedings of the Royal Society London B*, supp., 271, pp. S270–72.

Roberts, Tomi-Ann. 1991. Gender and the Influence of Evaluations on Self-Assessments in Achievement Settings. *Psychological Bulletin* 109, pp. 297–308.

Roberts, Tomi-Ann, and Susan Nolen-Hoeksema. 1989. Sex Differences in Reactions to Evaluative Feedback. *Sex Roles* 21, pp. 725–47.

Romer, David. 2006. Do Firms Maximize? Evidence from Professional Football. *Journal of Political Economy*, April, pp. 340–65.

Ross, Lee, Mark R. Lepper, and Michael Hubbard. 1975. Perseverance in Self-Perception and Social Perception: Biased Attributional Processes in the Debriefing Paradigm. *Journal of Personality and Social Psychology* 32 (5), pp. 880–92.

Rubin, David C. 1977. Very Long-Term Memory for Prose and Verse. *Journal of Verbal Learning and Verbal Behavior* 16, pp. 611–21.

———. 1994. *Memory in Oral Traditions*. New York: Oxford University Press.

Russo, J. Edward. 1977. The Value of Unit Price Information. *Journal of Marketing Research* 14, pp. 193–201.

Russo, J. Edward, and Paul J. H. Schoemaker. 1989. *Decision Traps*. New York: Doubleday.

———. 1992. Managing Overconfidence. *Sloan Management Review* 33, pp. 7–17.

Sabatini, Jeff. 2006. 2007 Mercedes-Benz S-Class: Leave the Driving to the Microchips. *New York Times*, May 28.

Salthouse, T. A. 1985. Anticipatory Processing in Transcription Typing. *Journal of Applied Psychology* 70, pp. 264–71.

Samuelson, William, and Richard Zeckhauser. 1988. Status Quo Bias in Decision Making. *Journal of Risk and Uncertainty* 1, pp. 7–59.

Sanna, Lawrence J., Norbert Schwarz, and Shevaun L. Stocker. 2002. When Debiasing Backfires: Accessible Content and Accessibility Experiences in Debiasing Hindsight. *Journal of Experimental Psychology: Learning, Memory, and Cognition* 28 (3), pp. 497–502.

Saul, Stephanie. 2006. Record Sales of Sleeping Pills Are Causing Worries. *New York Times*, Feb. 7.

Scharine, A., and M. McBeath. 2002. Right-Handers and Americans Favor Turning to the Right. *Human Factors* 44 (1), pp. 248–56.

Schendel, Joel D., John C. Morey, M. Janell Granier, and Sid Hall. 1983. Use of Self-Assessments in Estimating Levels of Skill Retention. U.S. Army Research Institute for the Behavioral and Social Sciences, Research Report 1341.

Schkade, David. 2007. Author interview.

Schkade, David, and Daniel Kahneman. 1998. Does Living in California Make People Happy? A Focusing Illusion in Judgments of Life Satisfaction. *Psychological Science* 9 (5), pp. 340–46.

Schoemaker, Paul J. H. 2007. Author interview.

Schoemaker, Paul J. H., and Robert E. Gunther. 2006. The Wisdom of Deliberate Mistakes. *Harvard Business Review*, June, pp. 109–15.

Schrage, Michael. 2003. Daniel Kahneman: The Thought Leader Interview. *Strategy + Business* (Winter), pp. 121–26.

Schwartz, Nelson D. 2006. One Brick at a Time. *Fortune*, June 6.

Searcey, Dionne. 2008. Generation Text: Emailing on the Go Sends Some Users into Harm's Way. *Wall Street Journal*, July 25, p. A1.

Senden, Marius von. 1960. *Space and Sight: The Perception of Space and Shape in the Congenitally Blind Before and After Operation.* London: Methuen.

Sexton, J. Bryan, Eric J. Thomas, and Robert L. Helmreich. 2000. Error, Stress, and Teamwork in Medicine and Aviation: Cross Sectional Surveys. *British Medical Journal* 320, pp. 745–49.

Shappell, Scott A., and Douglas A. Wiegmann. 2000. *The Human Factors Analysis and Classification System—HFACS.* Office of Aviation Medicine, Federal Aviation Administration, Report No. DOT/FAA/AM-00/7, Washington, D.C.

———. 2001. Unraveling the Mystery of General Aviation Controlled Flight into Terrain Accidents Using HFACS. *Proceedings of the 11th International Symposium on Aviation Psychology,* Columbus, Ohio.

———. 2003. *A Human Error Analysis of General Aviation Controlled Flight into Terrain Accidents Occurring Between 1990–1998.* Office of Aerospace Medicine, Federal Aviation Administration, Report No. DOT/FAA/AM-03/4, Washington, D.C.

Shepard, Roger N. 1990. *Mind Sights.* New York: W. H. Freeman.

Shepard, Roger N., and Lynn A. Cooper. 1982. *Mental Images and Their Transformations.* Cambridge, Mass.: MIT Press.

Shepherd, J., G. M. Davies, and H. D. Ellis. 1981. Studies of Cue Saliency. In *Perceiving and Remembering Faces,* edited by G. M. Davies, H. D. Ellis, and J. Shepherd. New York: Academic Press.

Shojania, Kaveh G., et al. 2003. Changes in Rates of Autopsy-Detected Diagnostic Errors over Time. *Journal of the American Medical Association* 289, pp. 2849–56.

Simon, William E. 1971. Number and Color Responses of Some College Students: Preliminary Evidence for a "Blue Seven Phenomenon." *Perceptual and Motor Skills* 33, pp. 373–74.

Simon, William E., and Louis H. Primavera. 1972. Investigation of the "Blue Seven Phenomenon" in Elementary and Junior High School Children. *Psychological Reports* 31, pp. 128–30.

Simons, Daniel J. 2007. Author interview.

Simons, Daniel J., and Daniel T. Levin. 1998. Failure to Detect Changes to People During Real-World Interaction. *Psychonomic Bulletin & Review* 5 (4), pp. 644–49.

Slamecka, Norman J. 1985. Ebbinghaus: Some Associations. *Journal of Experimental Psychology: Learning, Memory, and Cognition* 11 (3), pp. 414–35.

Sloboda, John. 1976. The Effect of Item Position on the Likelihood of Identification by Inference in Prose Reading and Music Reading. *Canadian Journal of Psychology* 30 (4), pp. 228–37.

———. 1985. *The Musical Mind.* Oxford: Clarendon Press.

———. 1988. *Generative Processes in Music.* Oxford: Clarendon Press.

———. 2005. *Exploring the Musical Mind.* Oxford: Oxford University Press.

Slovic, Paul. 1973. Behavioral Problems of Adhering to a Decision Policy. Paper presented at the spring seminar of the Institute for Quantitative Research in Finance, Napa, Calif.

Smith, Dylan M., George Loewenstein, Aleksandra Jankovich, and Peter A. Ubel. 2007. The Dark Side of Hope: Lack of Adaptation to Temporary Versus Permanent Colostomy. Unpublished manuscript.

Spangenberg, Eric R., et al. 2006. Effects of Gender-Congruent Ambient Scent on Approach and Avoidance Behaviors in a Retail Store. *Journal of Business Research* 59, pp. 1281–87.

Srinivasan, M. V., et al. 1996. Honeybee Navigation En Route to the Goal: Visual Flight Control and Odometry. *Journal of Experimental Biology* 199 (1), pp. 237–44.

Starkes, Janet L., and K. Anders Ericsson, eds. 2003. *Expert Performance in Sports: Advances in Research in Sport Expertise.* Champaign, Ill.: Human Kinetics.

Strack, Fritz, Leonard L. Martin, and Norbert Schwarz. 1988. Priming and Communication: Social Determinants of Information Use in Judgments of Life Satisfaction. *European Journal of Social Psychology* 18, pp. 429–42.

Strack, Fritz, and Thomas Mussweiler. 1997. Explaining the Enigmatic Anchoring Effect: Mechanisms of Selective Accessibility. *Journal of Personality and Social Psychology* 73 (3), pp. 437–46.

Strayer, David L., and Frank A. Drews. 2004. Profiles in Driver Distraction: Effects of Cell Phone Conversations on Younger and Older Drivers. *Human Factors* 46 (4), pp. 640–49.

Swensen, C. H. 1957. Empirical Evaluations of Human Figure Drawings. *Psychological Bulletin* 54, pp. 431–66.

Tan, Cheryl Lu-Lien. 2007. Hey, Honey Bunny, Stores Know What Your Wife Wants. *Wall Street Journal*, Dec. 1–2, p. A1.

Tat, Peter, William Cunningham, and Emin Babakus. 1988. Consumer Perception of Rebates. *Journal of Advertising Research*, Aug./Sept., pp. 45–49.

Tellegen, Auke, et al. 1988. Personality Similarity in Twins Reared Apart and Together. *Journal of Personality and Social Psychology* 54 (6), pp. 1031–39.

Tenney, Y. J. 1984. Aging and the Misplacing of Objects. *British Journal of Developmental Psychology* 2, pp. 43–50.

Tetlock, Philip E. 1998. Close-Call Counterfactuals and Belief-System Defenses: I Was Not Almost Wrong but I Was Almost Right. *Journal of Personality and Social Psychology* 75 (3), pp. 639–52.

———. 2005. *Expert Political Judgment: How Good Is It? How Can We Know?* Princeton, N.J.: Princeton University Press.

Thaler, Richard H., and Eric J. Johnson. 1990. Gambling with the House Money and Trying to Break Even: The Effects of Prior Outcomes on Risky Choice. *Management Science* 36 (6), pp. 643–60.

Thompson, William C., Geoffrey T. Fong, and D. L. Rosenhan. 1981. Inadmissible Evidence and Juror Verdicts. *Journal of Personality and Social Psychology* 40 (3), pp. 453–63.

Thurm, Scott, and Pui-Wing Tam. 2008. States Scooping Up Assets from Millions of Americans. *Wall Street Journal*, Feb. 4, p. A1.

Todorov, Alexander, et al. 2005. Inferences of Competence from Faces Predict Election Outcomes. *Science* 308, pp. 1623–26.

Tolman, E. C. 1948. Cognitive Maps in Rats and Men. *Psychological Review* 55, pp. 189–208.

TowerGroup. 2006. With Soaring Gift Cards Sales Poised to Exceed $80 Billion in 2006, Unused Card Values Are Also on the Rise. Press release, Nov. 20.

Tsimhoni, Omer, Daniel Smith, and Paul Green. 2004. Address Entry While Driving: Speech Recognition Versus a Touch-Screen Keyboard. *Human Factors* 46 (4), pp. 600–610.

Tugend, Alina. 2007. The Many Errors in Thinking About Mistakes. *New York Times*, Nov. 24.

Tulving, Endel, and Fergus I. M. Craik, eds. 2000. *The Oxford Handbook of Memory*. New York: Oxford University Press.

Tversky, Amos, and Daniel Kahneman. 1981. The Framing of Decisions and the Psychology of Choice. *Science* 211, pp. 453–58.

Tversky, Barbara. 1981. Distortions in Memory for Maps. *Cognitive Psychology* 13, pp. 407–33.

———. 2004. Narratives of Space, Time, and Life. *Mind & Language* 19 (4), pp. 380–92.

Tversky, Barbara, and Elizabeth Marsh. 2000. Biased Retellings of Events Yield Biased Memories. *Cognitive Psychology* 40, pp. 1–38.

U.S. Department of Health and Human Services. 2004. Almost Half of Americans Use at Least One Prescription Drug Annual Report on Nation's Health Shows. Press release, Dec. 2.

U.S. Government Accountability Office. 2007. *Older Driver Safety*. Report to the Special Committee on Aging, U.S. Senate. GAO-07-413, April.

Vallone, Robert P., et al. 1990. Overconfident Prediction of Future Actions and Outcomes by Self and Others. *Journal of Personality and Social Psychology* 58 (4), pp. 582–92.

Vander Molen, Tom. 2007. Author interview.

Vanhuele, Marc, Grilles Laurent, and Xavier Dreze. 2006. Consumers' Immediate Memory for Prices. *Journal of Consumer Research* 33 (2), pp. 163–72.

Vickers, Joan N. 1992. Gaze Control in Putting. *Perception* 21, pp. 117–32.

———. 1996. Visual Control When Aiming at a Far Target. *Journal of Experimental Psychology: Human Perception and Performance* 22 (2), pp. 342–54.

Vincent, Charles. 2003. The Other Side. *Morbidity & Mortality Rounds on the Web*, Oct.

Vlasic, Bill. 2008. More Options to Tempt Eyes Off Road, Hands Off Wheel. *New York Times*, Feb. 12, p. 1.

Waber, Rebecca L., Baba Shiv, Ziv Carmon, and Dan Ariely. 2008. Commercial Features of Placebo and Therapeutic Efficacy. *Journal of the American Medical Association* 299 (9), pp. 1016–17.

Wade, Elizabeth, and Herbert Clark. 1993. Reproduction and Demonstration in Quotations. *Journal of Memory and Language* 32, pp. 805–19.

Wald, Matthew L. 2007. Fatal Crashes of Airplanes Decline 65% over 10 Years. *New York Times*, Oct. 1, p. C1.

Wallace, Wanda T. 1994. Memory for Music: Effect of Melody on Recall of Text. *Journal of Experimental Psychology: Learning, Memory, and Cognition* 20, pp. 1471–85.

Wallace, Wanda T., and David C. Rubin. 1988a. Memory of a Ballad Singer. In *Practical Aspects of Memory: Current Research and Issues.* Vol. 1: *Memory in Everyday Life*, edited by M. M. Gruneberg, P. E. Morris, and R. N. Sykes. Chichester, U.K.: Wiley.

———. 1988b. Wreck of the Old 97: A Real Event Remembered in Song. In *Remembering Reconsidered: Ecological and Traditional Approaches to the Study of Memory*, edited by Ulric Neisser and Eugene Winograd. Cambridge, U.K.: Cambridge University Press.

Wang, Jing-Shiarn, Ronald R. Knipling, and Michael J. Goodman. 1996. The Role of Driver Inattention in Crashes: New Statistics from the 1995 Crashworthiness Data System. *40th Annual Proceedings, Association for the Advancement of Automotive Medicine*, Vancouver, British Columbia.

Wansink, Brian, Robert J. Kent, and Stephen J. Hoch. 1998. An Anchoring and Adjustment Model of Purchase Quantity Decisions. *Journal of Marketing Research* 35, pp. 71–81.

Watt, Christopher J. 2000. An Analysis of Decision Making Strategies Used by P-3 Pilots in Hazardous Situations. Master's thesis, Naval Postgraduate School, Monterey, Calif.

Weber, Elke U. 1997. Perception and Expectation of Climate Change. In *Environment, Ethics, and Behavior*, edited by Max H. Bazerman et al. San Francisco: Jossey-Bass.

————. 2007. Author interview.

Weber, Elke U., Ann-Renee Blais, and Nancy E. Betz. 2002. A Domain-Specific Risk-Attitude Scale: Measuring Risk Perceptions and Risk Behaviors. *Journal of Behavioral Decision Making* 15, pp. 263–90.

Weinstein, Neil D., and William M. Klein. 1995. Resistance of Personal Risk Perceptions to Debiasing Interventions. *Health Psychology* 14 (2), pp. 132–40.

Welsh, Jonathan. 2008. Eyes for Night Driving. *Wall Street Journal*, Feb. 21, p. D6.

Wilkinson, Julie. 1988. Context Effects in Children's Event Memory. In *Practical Aspects of Memory: Current Research and Issues*. Vol. 1: *Memory in Everyday Life*, edited by M. M. Gruneberg, P. E. Morris, and R. N. Sykes. Chichester, U.K.: Wiley.

Williamson, A. M., and Anne-Marie Feyer. 2000. Moderate Sleep Deprivation Impairments in Cognitive and Motor Performance Equivalent to Legally Prescribed Levels of Alcohol Intoxication. *Occupational and Environmental Medicine* 57, pp. 649–755.

Willis, Janine, and Alexander Todorov. 2006. First Impressions: Making Up Your Mind After a 100-Ms Exposure to a Face. *Psychological Science* 17, pp. 592–98.

Wilson, Craig. 2008. A Bumper Crop of Car Gizmos Boggles the Mind. *USA Today*, May 14, p. 1D.

Wilson, Timothy D., and Nancy Brekke. 1994. Mental Contamination and Mental Correction: Unwanted Influences on Judgments and Evaluations. *Psychological Bulletin* 116 (1), pp. 117–42.

Winograd, Eugene. 1976. Recognition Memory for Faces Following Nine Different Judgments. *Bulletin of the Psychonomic Society* 8, pp. 419–21.

Winograd, Eugene, and Robert M. Soloway. 1986. On Forgetting the Locations of Things Stored in Special Places. *Journal of Experimental Psychology: General* 115 (4), pp. 366–72.

Wohlstetter, Roberta. 1962. *Pearl Harbor: Warning and Decision.* Stanford, Calif.: Stanford University Press.

Wolf, E. 1967. Studies on the Shrinkage of the Visual Field with Age. *Transportation Research Record 164,* Transportation Research Board, National Academy of Sciences.

Wolf, Thomas. 1976. A Cognitive Model of Musical Sight-Reading. *Journal of Psycholinguistic Research* 5 (2), pp. 143–71.

Wolfe, Jeremy M. 2007. Author interview.

Wolfe, Jeremy M., Todd S. Horowitz, and Naomi M. Kenner. 2005. Rare Items Often Missed in Visual Searches. *Nature* 435, pp. 439–40.

Wu, Suzanne. 2007. Are You Ready for Professional-Grade Golf Clubs? Public release from University of Chicago Press Journals, May 10.

Wulf, Steve. 1992. He Is an Einstein: Joe Theismann's Quote Tests the Theory of Relativity. *Sports Illustrated,* March 2, p. 12.

Yarmey, A. D. 1973. I Recognize Your Face but Can't Remember Your Name: Further Evidence on the Tip-of-the-Tongue Phenomenon. *Memory & Cognition* 1, pp. 287–90.

Yates, Frances Amelia. 1966. *The Art of Memory.* Chicago: University of Chicago Press.

Young, Andrew W. 1993. Recognizing Friends and Acquaintances. In *Memory in Everyday Life,* edited by Graham M. Davies and Robert H. Logie. Amsterdam: North-Holland.

———. 1998. *Face and Mind.* Oxford: Oxford University Press.

Young, Andrew W., Dennis C. Hay, and Andrew W. Ellis. 1985. The Faces That Launched a Thousand Slips: Everyday Difficulties and Errors in Recognizing People. *British Journal of Psychology* 76, pp. 495–523.

Zambito, Thomas. 2007. Casino Klutz Sues in Picasso Slipup. *New York Daily News,* Jan. 12, p. 2.

# INDEX

## About the Author

Pulitzer Prize–winning journalist Joseph T. Hallinan
has spent years collecting and cataloguing human errors
and the reasons behind them. A former Nieman Fellow
at Harvard University, Hallinan was most recently a reporter for the *Wall Street Journal*. He lives in Chicago
with his wife, Pam, and their three children. This is his
second book.